Advances in Legume Research: Physiological Responses and Genetic Improvement for Stress Resistance

(Volume 2)

Edited by

Phetole Mangena

Department of Biodiversity
School of Molecular and Life Sciences
Faculty of Science and Agriculture
University of Limpopo, Limpopo Province
Republic of South Africa

&

Sifau A. Adejumo

Department of Crop Protection and Environmental Biology
Faculty of Agriculture
University of Ibadan, Ibadan,
Nigeria

Advances in Legume Research: Physiological Responses and Genetic Improvement for Stress Resistance

(Volume 2)

Editors: Phetole Mangena & Sifau A. Adejumo

ISSN (Online): 2737-4890

ISSN (Print): 2737-4882

ISBN (Online): 978-981-5165-31-9

ISBN (Print): 978-981-5165-32-6

ISBN (Paperback): 978-981-5165-33-3

Published by Bentham Science Publishers Pte. Ltd. Singapore. All Rights Reserved.

First published in 2023.

need for a court order if at any point you breach any terms of this License Agreement. In no event will any delay or failure by Bentham Science Publishers in enforcing your compliance with this License Agreement constitute a waiver of any of its rights.

3. You acknowledge that you have read this License Agreement, and agree to be bound by its terms and conditions. To the extent that any other terms and conditions presented on any website of Bentham Science Publishers conflict with, or are inconsistent with, the terms and conditions set out in this License Agreement, you acknowledge that the terms and conditions set out in this License Agreement shall prevail.

Bentham Science Publishers Pte. Ltd.
80 Robinson Road #02-00
Singapore 068898
Singapore
Email: subscriptions@benthamscience.net

BENTHAM SCIENCE

CONTENTS

FOREWORD

Legumes play important roles in human diets. They serve as the main source of proteins especially for resource-poor families. They represent important sources of human and animal feeds that are rich in protein. More importantly, by their symbiotic interactions with nitrogen-fixing bacteria, apart from their contribution to food security and nutrition, they play a major role in climate change mitigation by serving as an alternative and sustainable strategy for improving soil fertility. Though, the Green Revolution in agriculture has helped in meeting the demands for food security by developing new crop varieties and increasing the use of synthetic nitrogen (N) fertilizer that is also contributing to climate change,, legume production does not depend on the use of synthetic fertilizers. It rather improves soil nutrient status and reduces over-dependence on nitrogen fertilizers. Legumes, therefore, should be considered key components in a sustainable agronomic programme. Their production, however, faces many challenges which are grouped under biotic and abiotic stresses.

To take advantage of these potential benefits of legumes, there is a need for a thorough understanding of the challenges faced by farmers in growing leguminous crops. High up on the list of challenges are the threats posed by a range of biotic stresses. It is therefore of immense value that these stresses are so effectively described in this volume. It is a comprehensive and expansive consideration of how biotic stress impacts legumes and how they can be managed. Further, as the authors are all working in Africa, they offer a unique perspective on the potential of legumes in a continent that is witnessing a substantial increase in human population and where climate change is also a major concern. The United Nations estimates that the human population will reach 8 billion at the end of 2022, representing an increase of one billion new mouths to feed. Against this background, the increase in the production of leguminous crops offers obvious attractions. I, therefore, unequivocally recommend this book to agronomists and to general science readers.

Luis Mur
Director of Research: Biology and Health
Aberystwyth University, Aberystwyth
UK

PREFACE

When the first volume of **Advances in Legume Research** was published in 2020 it was not anticipated that this next volume would soon begin, with so much interest of our authors sparked by the previous one. As the previous volume reported a vast amount of advanced information regarding both biotic and abiotic stress-induced reductions in the growth and yield of legumes, particularly, cowpea, mung bean and soybean. Presently, we focussed the current volume on pertinent literature and specific new developments that belong to the topic, as chosen for this book. As earlier envisioned, this book is intended to share new developments pertaining to the ways in which biotic stress factors continue to inflict harm on leguminous crops, as they are among the most vulnerable and highly sensitive groups of oilseed crops worldwide. Although it is aimed at both experienced and newcomer researchers/ students, this book offers new insights for individuals looking for new perspectives in the current knowledge of diseases and pests, associated with legumes, as well as the mechanisms in which these crops may or may not resist these attacks. Mostly, the book focuses on the influences of bacteria, fungi, viruses, arthropodous spiders and invertebrate organisms, as well as how climate change drives the population diversity and distribution of these microbial pests in order to limit plant growth and productivity in leguminous crops.

Such a book is highly required, especially to grow our knowledge and understanding of how the genetic diversity of crop plants can be protected, improved and sustained to benefit the current and future generations. This endeavour can be beefed up by establishing efficiently analysed genomics and proteomics data that provide concrete insights underlining molecular mechanisms that play a critical role in enabling crops to effectively adapt and respond to biotic stress, as highlighted in the introductory chapter of this book. As we look ahead to the possible preparation of the next volume, we hope that readers of this and previous volumes will find time and space to provide us with critical comments, suggestions or errors, if any. Finally, we are very indebted to Dr. Mabila and Ms. Noko Monene (Department of Research Administration and Development, University of Limpopo, South Africa) for their continued financial support, Prof. Luis Mur for providing his expertise outlook and reasons why readers must read this book. Also, many thanks are due to our publisher for all the help we received and for patiently waiting for documents. We are especially grateful to the authors and everyone who assisted over the period of preparing this volume.

Phetole Mangena
Department of Biodiversity
School of Molecular and Life Sciences
Faculty of Science and Agriculture
University of Limpopo, Limpopo Province
Republic of South Africa

Sifau A. Adejumo
Department of Crop Protection and Environmental Biology
Faculty of Agriculture
University of Ibadan, Ibadan
Nigeria

ACKNOWLEDGEMENT

Yet again, the paramount goal of delivering a comprehensive book that clearly elucidates the understanding of physiological and genetic mechanisms to confer biotic stress resistance was made possible by all contributors. As such, we are very much grateful to all the authors and everyone who provided their high-quality contributions. We express our special thanks and appreciation to Bentham Science for the support and help in making this Volume 2 achievable.

Phetole Mangena
Department of Biodiversity
School of Molecular and Life Sciences
Faculty of Science and Agriculture
University of Limpopo, Limpopo Province
Republic of South Africa

Sifau A. Adejumo
Department of Crop Protection and Environmental Biology
Faculty of Agriculture
University of Ibadan, Ibadan
Nigeria

List of Contributors

Abideen A. Alarape	Department of Wildlife and Ecotourism Management, Faculty of Renewable Natural Resources, University of Ibadan, Ibadan, Nigeria
Arinao Mukatuni	Department of Chemical Sciences, Faculty of Science, University of Johannesburg, Doornfontein Campus, P. O. Box 17011, Johannesburg 2028, South Africa
Benjamin Joshua	Department of Crop Protection, Georg August University, Gottingen, Germany
Hafiz A. Badmus	Department of Biodiversity, School of Molecular and Life Sciences, Faculty of Science and Agriculture, University of Limpopo, Limpopo Province, Republic of South Africa Department of Crop Protection and Environmental Biology, Faculty of Agriculture, University of Ibadan, Ibadan, Nigeria
Ifedolapo O. Adebara	Department of Crop Protection and Environmental Biology, Faculty of Agriculture, University of Ibadan, Ibadan, Nigeria
Josephine Malatji	Department of Microbiology, Biochemistry and Biotechnology, School of Molecular and Life Sciences, Faculty of Science and Agriculture, University of Limpopo, Limpopo Province, Republic of South Africa
Mokgadi Asnath Modiba	Department of Biodiversity, School of Molecular and Life Sciences, Faculty of Science and Agriculture, University of Limpopo, Limpopo Province, Republic of South Africa
Phetole Mangena	Department of Biodiversity, School of Molecular and Life Sciences, Faculty of Science and Agriculture, University of Limpopo, Limpopo Province, Republic of South Africa
Phumzile Mkhize	Department of Microbiology, Biochemistry and Biotechnology, School of Molecular and Life Sciences, Faculty of Science and Agriculture, University of Limpopo, Limpopo Province, Republic of South Africa
Pirtunia Nyadzani Mushadu	Department of Biodiversity, School of Molecular and Life Sciences, Faculty of Science and Agriculture, University of Limpopo, Limpopo Province, Republic of South Africa
Sifau A. Adejumo	Department of Crop Protection and Environmental Biology, Faculty of Agriculture, University of Ibadan, Ibadan, Nigeria
Sinorita Chauke	Department of Biodiversity, School of Molecular and Life Sciences, Faculty of Science and Agriculture, University of Limpopo, Limpopo Province, Republic of South Africa
Yolette Belinda Rapelang Nyathi	Department of Biodiversity, School of Molecular and Life Sciences, Faculty of Science and Agriculture, University of Limpopo, Limpopo Province, Republic of South Africa

CHAPTER 1

Biotic Stress and Breeding of Plants for Stress Resistance

Phetole Mangena[1,*] and **Sifau A. Adejumo**[2]

[1] *Department of Biodiversity, School of Molecular and Life Sciences, Faculty of Science and Agriculture, University of Limpopo, Limpopo Province, Republic of South Africa*

[2] *Department of Crop Protection and Environmental Biology, Faculty of Agriculture, University of Ibadan, Ibadan, Nigeria*

Abstract: Among the different environmental challenges that affect crop production, biotic stress factors are more devastating. They reduce crop yield and pose serious threats to food security. Legumes constitute a large number of crop varieties that are seriously affected by different biotic stress factors. To enhance their growth in the face of these different stressful factors and preserve their useful genomic and functional growth properties, leguminous crops are subjected to continuous genetic manipulations for stress resistance. Successful breeding of stress-tolerant varieties for cultivation under different farming systems may result in reduced crop losses and production costs, limited use of agrochemicals, and eventual yield increases. Crops that are resistant to biotic stress also exhibit better growth and yield characteristics. As established several decades ago, the revolution in genomic research led to the development of many sophisticated and advanced crop improvement techniques that can be applied across a whole range of leguminous crop species such as cowpea, faba bean, lentil, mungbean, pea, soybean, *etc*. However, interest in genetic engineering, chemically-or-physically-based mutation breeding, marker-assisted selection, quantitative trait loci and genome editing (CRISPR-Cas) have expanded research beyond biotic stress resistance. These techniques play a key role in applications such as the manufacturing of bioenergy, and crop engineering for the expression of valuable bioactive compounds and recombinant proteins. This chapter briefly reviews the diversity of biotic stress factors (bacteria, fungi, insects, parasitic nematodes and viruses) and possible ways in which these stress factors can be managed and eradicated using various breeding methods. The review shows that the biotechnological tools mentioned above provide beneficial functions in pest management through genetic, physiological and morphological improvements, especially when coupled with other farming practices.

Keywords: Biotic stress, Genetic engineering, Resistance, Leguminous crops.

---

* **Corresponding author Phetole Mangena:** Department of Biodiversity, School of Molecular and Life Sciences, Faculty of Science and Agriculture, University of Limpopo, Limpopo Province, Republic of South Africa; Tel: +2715-268-4715; E-mail: phetole.mangena@ul.ac.za

DEFINING BIOTIC STRESS

Biotic stress can be broadly defined as any living component of the environment that prevents the plant from achieving its full genetic potential. Therefore, biotic stress refers to all negative influences caused by living organisms such as parasitic nematodes, viruses, disease-causing bacteria, fungi, arachnids, weeds, and insect pests. According to Gull *et al.* [1], biotic stresses reduce growth rates and cause major pre- and post-harvesting losses. The stress negatively influences the rate of photosynthesis as a result of reduction in leaf area, for instance, by insect pests. Microbial pathogens such as *Xanthomonas axonopodis* pv. *citri* also reduce photosynthesis by negatively affecting the activity of key enzymatic proteins such as Rubisco (ribulose 1,5 bisphosphate carboxylase), Rubisco activase and ATPase (Adenosine Triphosphate synthase) [11]. Taiz *et al.* [2] therefore referred to this kind of stress, including abiotic stress, as growth-inhibiting conditions that may not allow plants to achieve maximum growth and reproductive capacities. Legumes are one of the major groups of crop species serving as the most important components of both smallholder and large-scale farming systems across the tropical and subtropical regions and are severely affected by this kind of stress. These crops are predominantly cultivated in regions such as Asia, sub-Saharan Africa and Latin America where they serve as critical sources of good-quality dietary proteins, minerals, and oils.

The high value of legume grain seeds in promoting human and animal livelihoods, economic benefits and the improvement of soil quality (through the establishment of symbiotic relationship with nitrogen-fixing bacteria) led to several crop species being opted for cultivation as either monocrops or mixed cropping systems with cereals. However, they are more susceptible to different biotic stresses compared to other non-leguminous crops because of their proteinous nature. Their vegetative and yield characteristics, such as plant height, leaf/branch number, biomass, fruit and seed quantities are all affected by biotic stress. Some common microbial and insect pests that cause damage and diseases in legumes and other crops are summarised in Table 1. The table indicates some of the most common types of living organisms that co-exist with plants in their immediate environment. Although some of these organisms have mutually beneficial interactions with plants, others could be parasitic or pathogenic species and become detrimental to plant growth. These organisms include microbial pathogens like *Xanthomonas campestris pv. phaseoli*, *Fusarium oxysporum f.sp. ciceris*, *Leveillula taurica cv. Arn*, Alfalfa mosaic virus (AMV) and herbivorous insects like leafhoppers as well as beetles (Table 1), including the arthropods not indicated in the table.

Table 1. Some of the most common biotic stress factors negatively affecting leguminous crops under diverse environmental conditions.

Category	Species	Disease/ Common Name	References
Bacteria	*Pseudomonas syringae pv. phaseolicola*	Halo blight	Schwartz [3]
-	*Pseudomonas syringae pv. syringae*	Bacterial brown spot	
-	*Xanthomonas campestris pv. phaseoli*	Bacterial blight	
Fungi	*Fusarium oxysporum f.sp.ciceris*	Fusarium wilt	Hardaningsih [4]
-	*Fusarium solani*	Black root rot	
-	*Leveillula taurica cv. Arn*	Powdery mildew	
-	*Erysiphe spp.*	Powdery mildew	
-	*Uromyces cicer-arietini* [Gorgn.]	Rust	
-	*Rhizoctonia spp.*	Dry/wet root rots	
-	*Sclerotium rolfsii*	Collar rot	
Nematodes	*Meloidogyne spp.*	Root knot	Davis and Mitchum [5]
Viruses	*Alfalfa Mosaic Virus (AMV)*	-	Chatzivassiliou [6]
-	*Beet Western Yellow Virus (BWYV)*	-	
-	*Broad Bean Mosaic Virus (BBMV)*	-	
-	*Seed Borne Mosaic Virus (SBMV)*	-	
-	*Broad Bean Wilt Virus (BBWV)*	-	
-	*Bean Golden Mosaic Virus (BGMV)*	-	
Insects pests	*Empoasca spp.*	Leafhopper	Edwards and Singh [7] Singh and van Emden [8]
-	*Aphis craccivora*	Aphid	
-	*Ophiomyia phaseoli syn. Melanagromyza phaseoli*	Beanfly	
-	*Ootheca mutabilis,*	Beetle	
-	*Mylabris spp.*	Bettle	
-	*Medythia guaterna,*	Beetle	
-	*Nezara spp*	Bug	
-	*Anoplocnemis spp*	Bug	
-	*Riptortus spp.*	Bug	
-	*Acanthomia spp.*	Bug	

In response to biotic stress, plants have evolved intricate defense mechanisms to deal with the harmful effects of pests and microbial pathogens. These involve morphological, physiological, biochemical and molecular mechanisms. These are induced by plants in order to cope or deal with different biotic attacks and enhance crop productivity. According to Iqbal *et al.* [9], defense mechanisms can be triggered either when the toxic secondary metabolite released by the pathogen reaches the plant's internal cellular compartments or the system is activated upon immediate detection of the attack through inducible defense mechanisms which utilise specific detection and signal transduction pathways. Both specific detection systems and signal transduction mechanisms can help the plant sense the presence of an herbivore or pathogen and then alter its gene expression and metabolism accordingly to counter the stress.

Plants normally use mechanical barriers such as the cell wall (with silica in certain species), cuticle (a waxy outer layer), periderm or papillae formation as a constitutive defense system. The complex structure of papilla cells is formed between the cells' plasma membrane and the inside of the cell wall. Huckelhoven [10] referred to these cell appositions as the ones responsible for the prevention of fungal pathogens and blocking them from penetrating the plants' cell walls. This report also revealed that the molecular composition of these papillae differs significantly from those of the primary and secondary cell walls. Plants also use toxic secondary metabolites to defend themselves against insect pests and other herbivores. Further below, we survey literature and discuss the existing wide range of biotic stress factors, and the extent to which these organisms affect grain legume's growth and yield. Highlights on the diverse biotechnological mechanisms developed globally to help crop plants in overcoming biotic stress are also discussed. This chapter and the book as a whole at a more advanced level, elaborate on constitutive and inducible defenses, briefly take note of the beneficial interactions that exist amongst plants and microorganisms, and describe the role that other ecological factors (for example, climatic factors, physiographic factors, and animals) play in biotic stress evolution. The prevailing climate plays a key role in determining the type of stress factors that get imposed on crop plants, as well as the plant's ability to resist such attacks.

INSECT PESTS

Although legumes can successfully establish themselves under unfavourable conditions and with no fertilisation, as mentioned earlier, many of the species in this group suffer major losses in growth and yield due to insect pests. Singh and van Emden [8] suggested insect pests as the main probable cause of higher yield loses in legume production globally. A large number of different insects, covering numerous taxa attack all parts of the plant, from seed filling to maturation and assimilate storage, seedling development, vegetative growth, reproductive stages, harvesting, and beyond. As a result, a number of insect pests were found responsible for attacking legume plants and inducing stress. Most predominantly, insects are important pests of legumes because they damage plants through direct feeding and they also provide infection sites for microbial pathogens [5].

According to available reports, insects are also very important pests because they target different and most critical parts of the plant and at varying stages. Aphids (Fig. **1A, B**) for instance, occur at terminals of leaves and other plant parts, and suck plant sap, draining nutrients as well as vectoring viruses [11]. These insects feed on their host by inserting hypodermal needle-like stylet to reach the phloem sap. During this process, aphids secrete saliva-effector proteins such as diacetyl/L-xylulose reductase (DCXR) that disrupt the host's defense mechanisms. DCXR poses dual enzymatic functions in carbohydrate and dicarboxyl metabolism as produced by cowpea aphid, *Aphis craccivora* (Fig. **1C**) [12]. Examples of major insect pests of leguminous crops are shown in Fig. (**1**).

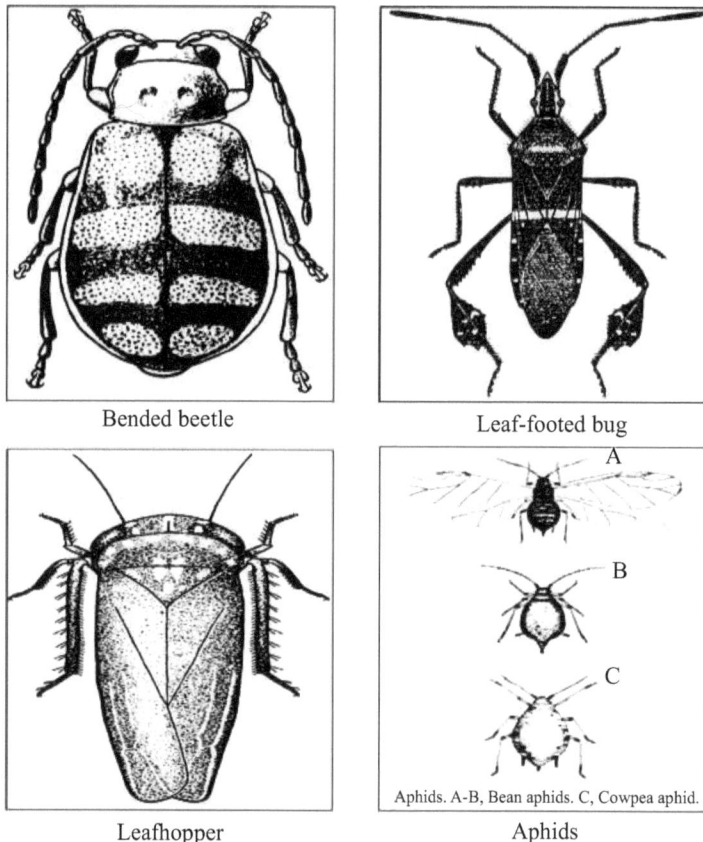

Bended beetle

Leaf-footed bug

Leafhopper

Aphids. A-B, Bean aphids. C, Cowpea aphid.

Aphids

Fig. (1). Major insect pests of leguminous crops, with aphids showing hypodermal needle-like stylets used to draw phloem sap [13, 14].

Singh and van Emden [8] further, presented a review of principal arthropodous pests of legume crops, grouped taxonomically in the order in which they colonise the crops. Among these pests, leafhoppers, aphids, bean-flies, beetles, Lepidoptera, thrips and pod-sucking *Hemiptera* were found to be the most widely spread pests. Arthropods such as spiders also play a key role in many agroecosystems of legumes by regulating pest species and serving as potential biological control agents. The phenotype of some of these insects is shown in Fig. (**1**) above. Hemiptera whitefly insects such as *Bemisia tabaci* have also been found to threaten the productivity of many crops, including grain legumes. Whitefly biotype B is considered a major pest for common bean (*Phaseolus vulgaris* L.). Similar to aphids, nymphs and adult whiteflies also cause direct damage to legumes by sucking phloem nutrients and inoculating salivary toxic enzymes [15].

The Cowpea weevil or cowpea seed beetle (*Callosobrachus maculatus*) was also reported as a major pest of cowpea, mungbean (*Vigna radiata*) and lentil (*Lens culinaris*). Beetles affect plant crops during the fruiting and post-harvest stage, leaving holes on the surface of pods and seeds. Generally, beetles and leafhoppers (Fig. **1**) serve as insect pests causing minor damage to legumes and other crops such as tomatoes and potatoes in the Solanaceae family. Insects do not only cause physical damage and phloem nutrient depletion in plants but, also serve as transmitting vectors of viruses (for example, BGMV and BBMV), meanwhile, caterpillars and leafhoppers do their best to only feed on leaves [13, 16]. Quresh *et al.* [13] further reported that thrips feed on blossoms, stink bugs, cornworms and leaf-footed bugs on seeds and fruit pods.

MICROBIAL PATHOGENS

It is likely that symbiotic associations between legumes and nitrogen-fixing microbes remain the best mutualistic relationship so far reported among these unrelated living organisms. This relationship that has been long in existence suggests a strong close co-evolution phenomenon between plants and endophytic/ mycorrhizal fungi as well as bacteria in the form of biofilms on the surfaces of roots and leaves. Endophytic bacteria and nitrogen-fixing bacteria are housed in specialised organs developed on the host plants, called nodules [2]. However, when clearly categorised, it was found that three major forms of beneficial interactions reported include associations with (i) rhizobacteria, (ii) mycorrhizal fungi and to a larger extent, (iii) nitrogen-fixing bacteria. These microbes provide a wide range of inorganic nutrients to plants, while in return, plants provide the microbes with carbohydrates and other metabolites.

Extensive literature is freely available that discusses how the beneficial interaction between microbes and plants works. In contrast, there are equally a group of microorganisms that cause infectious diseases in plants which also include fungi, molds, bacteria and viruses. Some of these common microbial pathogens affecting legume crops are highlighted in Table **1**. However, Taiz *et al.* [2] indicated that the majority of pathogens implicated for causing disease in Fabaceae species are fungi, belonging to Ascomycetes and Basidiomycetes. Meanwhile, disease-causing bacteria have been classified into three families, namely Xanthomonadaceae, Enterobacteriaceae and Pseudomonadaceae, which together constitute about 20 genera (Acidovorax, Erwinia, Brenneria, Burkholderia, Clavibacter, Candidatus, Dickeya, Erwinia, Liberibacter, Lonsdale, Pantoea, Pectobacterium, Phytoplasma, Pseudomonas, Ralstonia, Spiroplasma, Streptomyces, Xylella, Xanthomonas and Xylophilus) [17].

Gram-negative bacterium, *Pseudomonas syringae* pathovars remains the most common and well-studied bacterial pathogen. Disease symptoms caused by this bacterium include blights, stem cankers, leaf pots, and wilting as reported by Tripathi [18]. This species and other bacterial pathogens have been reported to cause serious damage to crop growth and productivity, even though most bacterial pathogens are considered to be less harmful and asymptomatic for almost the plant's entire life cycle. Viruses/ viroids have also been implicated in causing infectious diseases in plants. These organisms pose a serious challenge to disease control since the number of legume viruses continues to escalate [4]. According to Chatzivassiliou [6], the list of well-known viruses that infect legume crops currently stands at 168, with 39 and 16 genera as well as families, respectively. Some viruses become very important in causing serious diseases and major yield losses, especially for cool-season grain legume crops [19]. Recently emerged overwhelming evidence indicated that the use of winter grains, particularly those well adapted to spring season weather patterns may serve as the most effective way to increase crop production by minimising detrimental viruses that most commonly occur during cool seasons.

WEED PLANTS

Another type of biotic stress that is considered to be an undesirable economic pest in agricultural fields is the weed plant as described by Radosevich *et al.* [20]. According to Radosevich *et al.* [20], weed plants are said to have evolved from unintended consequences of cultivated crops. These weeds are highly persistent and competitive plants that seriously interfere negatively with the cultivation of all types of crops, including grain legumes and vegetables. Weed plants generally exhibit undesirable qualities that outweigh their good ecological functions. Some of these unwanted growth characteristics that make them more persistent are summarised in Table **2** below. According to the report of Radosevich *et al.* [20], weed plants can be considered as either anthropomorphic or biological with distinct and selective characteristics (Table **2**). Examples of some of the worst annual weed species reported by Tamms *et al.* [21] include *Amaranthus hybridus*, *A. spinosus*, *Avena fatua*, *Cynodon dactylon*, *Cyperus esculentus*, *Digitaria sanguinalis*, *Echinochloa colonum*, *E. crussipes* and *Sorghum halepense*. Some of these species evolved with a narrow adaptation to a single crop or cultivar [20]. Among the legumes, crops such as peas, faba beans, chickpea and lentils are mostly out-competed due to their small initial growth rates [22].

Table 2. Undesirable growth and reproductive characteristics of weed plants (1-5) and some of the problems that they cause in agriculture [6-10].

1. Abundant seed production and great seed longevity.
2. Rapid germination, vegetative and reproductive growth phases for population establishment.
3. Effective cross-pollination and adaptation for short-distance and long-distance dispersal of seeds and pollen.
4. Vigorous vegetative propagation and sexual reproduction.
5. Ability to compete inter-specifically using special traits such as rosettes and allelochemicals.
6. Reduce crop yield by competing for water, light, nutrients and space.
7. Use allelopathy (production of chemical substances which are toxic) against crop plants growing in the same environment.
8. Cause limitations on the choice of rotation sequence and cultural practices.
9. Serve as hosts for crop diseases and provide shelter for insect pests.
10. Interferes with harvest and reduces crop quality by contaminating the commodity.
Sources: Radosevich *et al* [20], Patches *et al.* [23]

BREEDING FOR BIOTIC STRESS RESISTANCE

Biological pests are an important limiting factor for the growth, yield and quality of grains produced by both legume and cereal crops. Biotic stress increasingly causes detrimental effects on many ecological, health and nutritional services derived from these agronomic crops, especially if the stress is allowed to persist. Herbicides, insecticides, bactericides and fungicides are frequently used to manage problems induced by pests in highly developed agricultural systems. However, the incorporation of these agrochemicals in farming practices can be constrained by production and input costs apart from their environmental implications, and increased development of insecticide-resistant pests such as the emergence of superbugs [22]. Zhang *et al.* [24] reported an intriguing whitefly priming phenomenon where whiteflies that first perceived herbivore-induced plant volatiles (HIPVs) modified and enhanced their suitability for their offspring to attack neighbouring host plants. HIPVs serve as a direct defense and repellent effect on plants to the insect herbivores. Upon perception of certain HIPVs, neighbouring pests prepare for plant attack and respond more rapidly as well as strongly with an appropriate defense reaction.

However, given these evolutional developments and the lack of capacity for farmers to deal with increasing biotic stress challenges, breeding methods must be advanced. For instance, farmers in developing regions such as East Asia, sub-Saharan Africa and South America are largely populated by small holdings that often face severe financial constraints, poor infrastructure, general lack of government support, and poor access to markets where they can sell their harvested products to grow their businesses. Limitations also exist involving the cultivation of legumes only in marginalised areas as a result of escalated input demands like the cost of application of herbicides, insecticides, microbicides, and new seeds. Therefore, modern biotechnological applications could be used to significantly reduce production costs as well as the use of agricultural chemicals that are aimed at achieving higher crop yields but could not guarantee a safe environment and less pollution. According to Zhang *et al.* [24], agrochemicals are often applied based on their potential economic benefits without evaluating their environmental impact.

Thus, breeding techniques such as genetic engineering, genome editing, mutation breeding, and marker-assisted selection serve as safer alternatives. Genome editing has currently emerged as one of the new efficient tools due to its ability to edit the genome of plants, together with microbial, animal, and human genomes [25]. Editing tools such as endonucleases and clusters of regularly interspaced short palindromic repeats (CRISP) and CRISPR-associated proteins (Cas) revealed the capability to unlock and alter DNA sequences to modify gene functions. Genome editing using transcription activator-like effectors (TALES) nucleases and zinc fingers (ZFs) reported several successes in the genetic manipulation of crops such as maize, cotton, rice, rapeseed, tobacco, and soybean for herbicide resistance [26]. But more work still needs to be done using this tool for legume crops known to be highly pest-susceptible (cowpea, peas, faba beans, and mungbean).

CONCLUDING REMARKS

According to ample already published reports, biotic stress factors like aphids, bacteria and viruses rapidly evolve mechanisms to detoxify pesticides [27]. For example, Aphids use effector proteins and aphid detoxification enzymes through the overexpression of DCXR to generate extra energy sources for their infestations and to alter plant defence response [28]. The use of pesticides to protect crop plants against these biotic stresses is a key element in agriculture. However, the application of agrochemicals is also criticised for its potential impact on the health of consumers, animals and the environment. Concerns raised range from the impact of these agricultural inputs on pollutant formation (soil, water, and air pollution) to transfer processes, including the gradual efforts made on the implementation and complementation of reduction strategies. Furthermore, conventional breeding methods (backcrossing, mass selection, pedigree, *etc.*), genetic engineering (direct and indirect-vector mediated transformation) and mutation breeding (physically and chemically induced DNA alterations) are still being investigated to improve crops against biotic stress factors.

Although the methods have been used for decades, others like the genetic transformation that is used in the production of genetically modified foods continue to receive a backlash from GMO opposing movements due to the undocumented potential safety concerns. Similarly, traditional breeding methods also face a number of specific limitations that include longer breeding cycles, biotypic variations, undesirable genetic linkages, lack of inter-specific cross-compatibility and poor germplasm [29]. These challenges suggest the need to develop more durable plant disease resistance methods as proposed by Kamphuis *et al.* [27], including identifying major resistance genes and quantitative trait loci, particularly, to confer insect pest resistance in legumes. Therefore, the development of new breeding protocols or integration of already existing methods remains critically important for a pest management strategy, agricultural sustainability, food security, and environmental protection.

LIST OF ABBREVIATIONS

AMV   Alfalfa Mosaic Virus

ATPase   Adenosine triphosphate synthase

BBMV   Broad Bean Mosaic Virus

BBWV   Broad Bean Wilt Virus

BGMV   Bean Golden Mosaic Virus

BWYV   Beet Western Yellow Virus

Cas   CRISPR-associated proteins

CRISPR Clusters of regularly interspaced short palindromic repeats

DCXR   Diacetyl-L-xylulose reductase

DNA   Deoxyribose nucleic acid

GMO   Genetically Modified Organisms

HIPVs   Herbivore-Induced Plant Volatiles

Rubisco Ribulose-1,5-bisphosphate carboxylase

TALES   Transcription Activator-Like Effector Nucleases

ZFs   Zinc Fingers

CONSENT FOR PUBLICATION

Not applicable.

CONFLICT OF INTEREST

The author declares no conflict of interest, financial or otherwise.

ACKNOWLEDGEMENT

Declared none.

REFERENCES

[1]   Gull, A.; Lone, A.A.; Wani, N.U. Biotic and abiotic stress in plants. In: *Abiotic and biotic stress in plants*; de Oliveira, A.B., Ed.; IntechOpen: London, **2019**; pp. 1-7.
[http://dx.doi.org/10.5772/intechopen.85832]

[2]   Taiz, L.; Zeiger, E.; Moller, I.M.; Murphy, M. *Plant physiology and development*; Sinauer Associates: London, **2015**, pp. 756-760.

[3]   Schwartz, HF *Bacterial diseases of beans*; Crop Series DiseasesColorado State University Extension, **2011**, 2. 913, pp. 1-3.

[4]   Hardaningsih, S. diseases of legume crops caused by fungi and their control. *BALITKABI,* **1999**, 92-98.

[5]   Davis, E.L.; Mitchum, M.G. Nematodes. Sophisticated parasites of legumes. *Plant Physiol.,* **2005**, *137*(4), 1182-1188.
[http://dx.doi.org/10.1104/pp.104.054973] [PMID: 15824280]

[6]   Chatzivassiliou, E.K. An annotated list of legume-infecting viruses in the light of metagenomics. *Plants,* **2021**, *10*(7), 1413.
[http://dx.doi.org/10.3390/plants10071413] [PMID: 34371616]

[7]   Edwards, O.; Singh, K.B. Resistance to insect pests: What do legumes have to offer? *Euphytica,* **2006**, *147*(1-2), 273-285.
[http://dx.doi.org/10.1007/s10681-006-3608-1]

[8]   Singh, S.R.; Emden, H F V. Insect pests of grain legumes. *Annu. Rev. Entomol.,* **1979**, *24*(1), 255-278.
[http://dx.doi.org/10.1146/annurev.en.24.010179.001351]

[9]   Iqbal, Z.; Iqbal, M.S.; Hashem, A.; Abd Allah, E.F.; Ansari, M.I. Plant defense response to biotic stress and its interplay with fluctuating dark/light conditions. *Front. Plant Sci.,* **2021**, *12*, 631810.
[http://dx.doi.org/10.3389/fpls.2021.631810] [PMID: 33763093]

[10]   Hückelhoven, R. The effective papilla hypothesis. *New Phytol.,* **2014**, *204*(3), 438-440.
[http://dx.doi.org/10.1111/nph.13026] [PMID: 25312607]

[11]   Garavaglia, B.S.; Thomas, L.; Gottig, N.; Zimaro, T.; Garofalo, C.G.; Gehring, C.; Ottado, J. Shedding light on the role of photosynthesis in pathogen colonization and host defense. *Commun. Integr. Biol.,* **2010**, *3*(4), 382-384.
[http://dx.doi.org/10.4161/cib.3.4.12029] [PMID: 20798833]

[12]   MacWilliams, J.R.; Dingwall, S.; Chesnais, Q.; Sugio, A.; Kaloshian, I. AcDCXR is a cowpea aphid effector with putative roles in altering host immunity and physiology. *Front. Plant Sci.,* **2020**, *11*, 605.
[http://dx.doi.org/10.3389/fpls.2020.00605] [PMID: 32499809]

[13]   Qureshi, JA; Seal, D; Webb, SE *Insect management for legumes (Beans, Peas)*; IFAS Ext;, **2021**, 465, pp. 1-14.
[http://dx.doi.org/10.32475/edis-ig151-20.3]

[14]   Sorensen, K.; Baker, J.; Cater, C.C.; Stephen, D. Pests of beans and peas. *NC State Extension Publications,* **2003**, *AG-235*, 1-5.

[15]   da Silva, A.G.; Boica Jr, A.L.; da Silva Farias, P.R.; de Souza, B.H.S.; Rodriques, N.E.L.; Carbonell, S.A.M. Common bean resistance expression to whitefly in winter and rainy seasons in Brazil. Sci. Agric. (Piracicada, Braz) *Entomology • Sci. agric. (Piracicaba, Braz.),* **2019**, *76*(5).
[http://dx.doi.org/10.32475/edis-ig151-20.3]

[16]   Onyido, A.E.; Zeibe, C.C.; Okonkwo, N.J.; Ezugbo-Nwobi, I.K.; Egbuche, C.M.; Udemezue, I.O.; Ezeanya, L.C. Damage caused by the bean bruchid, Callosobruchus maculatus (Fabricius) on different legume seeds in sale in Awka and Onitsha markets, Anambra State, Sout Easten Nigeria. *African Research Review,* **2011**, *5*(4), 116-123.
[http://dx.doi.org/10.4314/afrrev.v5i4.69264]

[17]   Agrios G. Plant pathology, 5th Eds. Gainesville, USA: Academic Press; pp. 615-703.

[18]   Tripathi, D. Bacterial pathogens in plants. *J. Bacteriol Mycol Open Access,* **2017**, *4*(2), 38-39.
[http://dx.doi.org/10.15406/jbmoa.2017.04.00083]

[19]   Makkouk, K.M.; Kumari, S.G.; van Leur, J.A.G.; Jones, R.A.C. Control of plant virus diseases in cool-season grain legume crops. *Adv. Virus Res.,* **2014**, *90*, 207-253.
[http://dx.doi.org/10.1016/B978-0-12-801246-8.00004-4] [PMID: 25410103]

[20]   Radosevich, S.R.; Holt, J.S.; Ghersa, C.M. Weeds and invasive plants. In: *Ecology of weeds and invasive plants: Relationship to agriculture and natural research management*; Radosevich, S.R.; Holt, J.S.; Ghersa, C.M., Eds.; John Wiley and Sons, Inc.: New Jersey, **2007**; p. 210.

[21]   Tamms, L.; de Mol, F.; Glemnitz, M.; Gerowitt, B. Weed densities in perennial flower mixtures cropped for greater arable biodiversity. *Agriculture,* **2021**, *11*(6), 501.
[http://dx.doi.org/10.3390/agriculture11060501]

[22]   Knott, C.M.; Halila, H.M. Weeds in food legumes: Problems, effects and control. In: *World crops: Cool season food legumes- current plant science and biotechnology in agriculture*; Summerfield, R.J., Ed.; Springer: Dordrecht, **1988**; pp. 535-548.
[http://dx.doi.org/10.1007/978-94-009-2764-3_45]

[23]   Patches, K.M.; Curran, W.S.; Lingenfelter, D.D. Effectiveness of herbicides for control of common pokeweed (Phytolacca americana) in corn and soybean. *Weed Technol.,* **2017**, *31*(2), 193-201.
[http://dx.doi.org/10.1614/WT-D-16-00043.1]

[24]   Zhang, L.; Yan, C.; Guo, Q.; Zhang, J.; Ruiz-Menjivar, J. The impact of agricultural chemical inputs on environment: Global evidence from informetrics analysis and visualization. *Int. J. Low Carbon Technol.,* **2018**, *13*(4), 338-352.
[http://dx.doi.org/10.1093/ijlct/cty039]

[25]   Tyagi, S.; Kesiraju, K.; Saakre, M.; Rathinam, M.; Raman, V.; Pattanayak, D.; Sreevathsa, R. Genome edifiting for resistance to insect pests: An emerging tool for crop improvement. *ACS Omega,* **2020**, *5*(33), 20674-20683.
[http://dx.doi.org/10.1021/acsomega.0c01435] [PMID: 32875201]

[26]   Makarova, K.S.; Koonin, E.V. Annotation and classification of CRISPR-Cas systems. *Methods Mol. Biol.,* **2015**, *1311*, 47-75.
[http://dx.doi.org/10.1007/978-1-4939-2687-9_4] [PMID: 25981466]

[27]   Kamphuis, L.G.; Zulak, K.; Gao, L.L.; Anderson, J.; Singh, K.B. Plant–aphid interactions with a focus on legumes. *Funct. Plant Biol.,* **2013**, *40*(12), 1271-1284.
[http://dx.doi.org/10.1071/FP13090] [PMID: 32481194]

[28]   Zhang, P.J.; Wei, J.N.; Zhao, C.; Zhang, Y.F.; Li, C.Y.; Liu, S-S.; Dicke, M.; Yu, X-P.; Turlings, T.C.J. Airborne host–plant manipulation by whiteflies *via* an inducible blend of plant volatiles. *Proc. Natl. Acad. Sci.,* **2019**, *116*(15), 7387-7396.
[http://dx.doi.org/10.1073/pnas.1818599116]

[29]   Keneni, G.; Bekele, E.; Getu, E.; Imtiaz, M.; Damte, T.; Mulatu, B.; Dagne, K. Breeding food legumes for resistance to storage insect pest: Potential and limitations. *Sustainability,* **2011**, *3*(9), 1399-1415.
[http://dx.doi.org/10.3390/su3091399]

CHAPTER 2

Current Knowledge on Biotic Stresses affecting Legumes: Perspectives in Cowpea and Soybean

Benjamin Joshua[1,*]

[1] *Department of Crop Protection, Georg August University, Gottingen, Germany*

Abstract: Legumes are economically important crops for the achievement of food security status in many countries in the tropical and subtropical regions of the world. Among various environmental stresses, biotic constraints to the production of grain legumes such as cowpea and soybean are becoming increasingly significant with the recurring change in climatic patterns and diverse environmental alterations. The economic impact of biotic factors such as disease-causing pathogens (fungi, bacteria, viruses and nematodes), insect pests and parasitic weeds has become overwhelming. These biotic stressors induce a wide range of damage symptoms which include stunting, wilting of stems, defoliation, root rots and premature death of plants. Yield losses due to the activities of biotic stress factors have been very significant. Hence, it is imperative to be informed of the various biotic stressors that affect the growth and yield potential of cowpeas and soybeans in various cropping systems. This review seeks to highlight existing pests and diseases in cowpea and soybean and evaluate their impact on the growth and productivity of these crops. It is hoped that the review will further spur scientific research into how these biotic factors can be managed or even manipulated to ensure agricultural sustainability, high economic returns, and global food security.

Keywords: Biotic Stressors, Cowpea, Diseases, Environmental Stress, Legumes, Food Security, Pests, Soybean.

INTRODUCTION

Leguminous crops belonging to the family Fabaceae are considered the most important grain crops after the grass or Gramineae family (*Poaceae*) [1]. Seeds of legumes are broadly used as direct food sources due to their high nutritional content and the presence of bioactive compounds such as flavonoids and polyphenols as well as micronutrients like essential vitamins and minerals [2, 3]. Amongst the leguminous crops, grain legumes such as peanuts, soybeans, dry

* **Corresponding author Benjamin Joshua:** Department of Crop Protection, Georg August University, Gottingen, Germany; Tel: +4915752071661; E-mail: benjaminjoshua1997@gmail.com

Phetole Mangena & Sifau A. Adejumo (Eds.)

beans, cowpeas and chickpeas, are considered key components of the human diet as they serve as the major supply of proteins [3]. The economic importance of grain legumes cannot be underemphasized as their mean annual global production from 2008 to 2017 is estimated at over 75 million tonnes [4]. About 14.5% of the global arable cropped area was occupied by grain legumes in 2014 [5].

Grain legumes are also critical sources of plant nutrients in a cereal production system as they possess the ability to incorporate biological nitrogen into the soil [6]. Hence they are crucial to the food security status of many regions of the world especially Africa. Like many other important food crops, legumes are vulnerable to different environmental stresses which could be abiotic or biotic [7, 8]. Abiotic stresses affecting legumes include drought, salinity, heat, high light intensity and nutrient imbalance [9], while major drivers of biotic stress include viruses, fungi, bacteria, nematodes, weeds and other parasites [10]. The occurrence of any stress conditions certainly affects the yield potential of legumes [8]. The composition and quality of grain legumes are negatively impacted by abiotic and biotic stresses [11].

These stress factors can influence the yields of legumes and other beneficiary cereals within a crop production system by inhibiting or promoting nodulation [12]. The response of legumes, like other crops, to environmental stresses varies based on the type of stress (biotic or abiotic), stress severity and plant vigour [13]. Although research shows that abiotic stress factors are known to impact legume production extensively [8, 11, 14, 15], biotic stresses are becoming more frequent owing to global warming and climate abnormalities [16, 17]. Hence, current knowledge of various biotic stress conditions affecting legumes, especially grains, will be more insightful.

BIOTIC STRESSES IN LEGUMES

Generally, legume growth and development are inhibited by many kinds of biotic stresses that induce direct and indirect physiological alterations [13]. These stress agents directly induce a deprivation of nutrients required by the host crop and can lead to the death of plants. High severity of biotic stress can bring about heavy pre- and postharvest losses [15]. Predominantly, the extent of damage influenced by biotic factors on legumes is highly dependent on the prevalence of one or more abiotic factors [9]. However, the major biotic stressors that can drastically reduce the yield of grain legumes predominantly involve microbial pathogens, pests, and weeds [18]. Therefore, the economic significance of the different biotic factors affecting the two major food legumes; cowpea and soybean, is then discussed below.

COWPEA

Cowpea (*Vigna unguiculata* L. Walp.) is arguably the most widely adapted, versatile and nutritious grain legume for both warm and dry agro-ecologies of the tropics and subtropics [19]. Cowpea belongs to the family Fabaceae and it is often called black-eye pea, southern pea or crowder pea. It is predominantly unique as a self-fertilizing crop [20]. Through all stages of development, it grows in a wide range of temperature from 18 to 28 °C. In Sub-Saharan Africa, it is widely grown for food and feed because its grain contains high proportion of protein (23 to 32%), energy, micro- and macro-nutrients [21, 22]. Being tolerant to harsh conditions, it is considered a valuable component in crop production systems of poor rural households [23]. Owing to its atmospheric nitrogen fixation ability, it readily serves as a crop for rotation with major cereals crops which are the main determinants of food security in developing countries [24]. Global production of cowpea in 2019 was over 8.9 million metric tonnes of which Africa accounted for over 97% [25, 26] (Fig. **1**) (Table **1**).

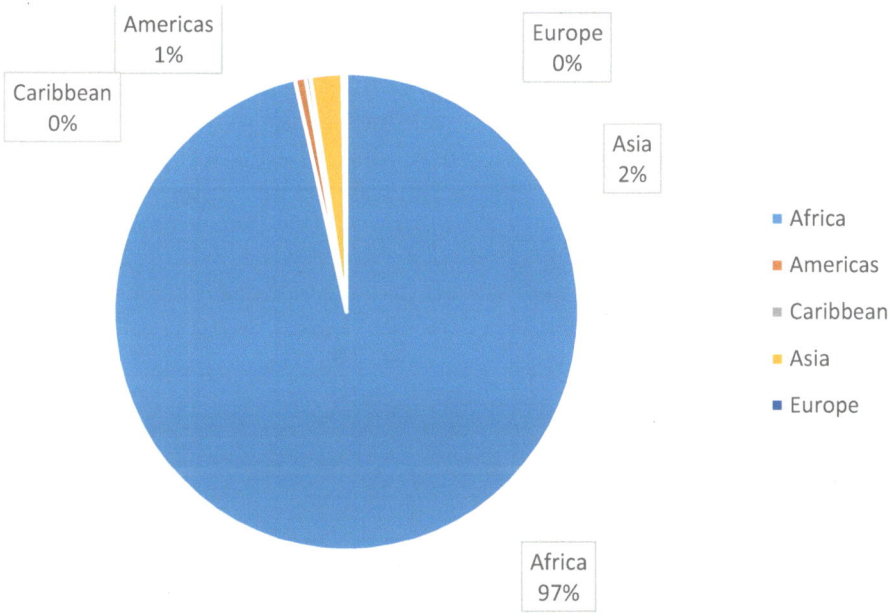

Fig. (1). Production of cowpea based on different regions of the world. FAOSTAT [26].

Table 1. Top 10 cowpea producing countries in the world (2019)..

Rank	Country	Production (Tonnes)	Yield (Kilogram/Hectare)	Area (Hectares)
1.	Nigeria	3576361	831.1	4303005
2.	Niger	2386735	416.9	5725433
3.	Burkina Faso	652454	481.8	1354100
4.	Ethiopia	374332	1701.2	220037
5.	Kenya	246870	828.1	298120
6.	Mali	215436	474.2	454274
7.	Cameroon	215016	881.0	244058
8.	Ghana	202735	1359.7	149102
9.	Senegal	184137	633.5	290677
10.	Sudan	161000	473.8	339780

Source: FAOSTAT (2019).

However, cowpea's productivity under traditional systems is considerably low especially in west Africa sub-region (0.025-0.3 tonnes/ha). This is often associated with drought and a wide array of biotic and other abiotic stresses [27, 28]. IITA [28] cowpea breeding programmes indicated that this crop is very sensitive to biotic stressors, particularly disease-causing pathogens (viruses, bacteria, and fungi), nematodes, insects, and weeds which induce extensive yield reductions [22] (Table **2**).

Table 2. Nutrient profile of cowpea mature seed (per 100 g edible portion on fresh weight basis).

Energy (kJ)	Protein (g)	Fat (g)	Ash (g)	Dietary Fibre (g)	K (mg)	P (mg)	Mg (mg)
1440	23.8	2.07	3.39	10.7	1380	438	333

Source: USDA, 2018 [27].

SOYBEAN

Soybean, (*Glycine max* (L.) Merr.), is an important seed legume produced in over 70 countries. It is cultivated in many regions of the world as a major source of protein and oil for livestock feed formulation and human consumption [29, 30]. It contributes to 25% of the global edible oil which is about two-thirds of the world's protein concentrate for livestock feeding [31]. Soybean is also crucial in the cropping systems of Africa owing to its high atmospheric nitrogen fixing potential [32]. About 6% of the world's arable land, an estimated total area of more than 92.5 million hectares, is used to produce soybeans [30]. Global production of soybean is estimated to be over 333 million tonnes per annum in

2019 with Brazil being the leading producer, followed by the United States and Argentina [26] (Fig. **2**) (Tables **3** & **4**).

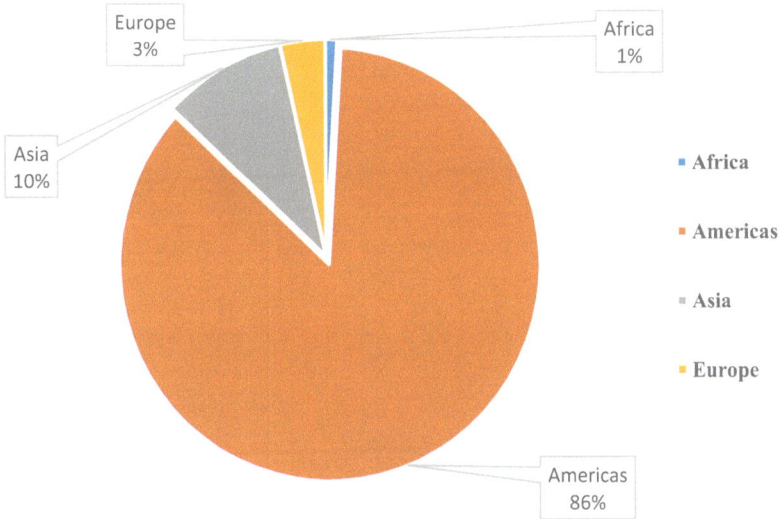

Fig. (2). Production of soybean based on different regions of the world. FAOSTAT [26].

Table 3. Nutrient profile of soybean mature seed (per 100 g edible portion on fresh weight basis).

Energy (kJ)	Protein (g)	Fat (g)	Ash (g)	Dietary Fibre (g)	K (mg)	P (mg)	Mg (mg)
1870	36.5	19.9	4.87	9.3	1800	704	280

Source: USDA, 2018 [27].

Table 4. Top 12 soybean producing countries in the world (2019).

Rank	Countries	Production (Kilo Tonnes)	Yield (Kg/Hectare)	Area (Hectare)
1.	Brazil	114269	3588.1	31846
2.	USA	96793	3035.2	31890
3.	Argentina	55263	1657.5	33340
4.	China, mainland	15724	842.3	18668
5.	India	13267	1113.1	11919
6.	Paraguay	8520	356.5	23900
7.	Canada	6045	227.0	26625
8.	Russia	4359	277.2	15727
9.	Ukraine	3698	161.2	22933

(Table 4) cont.....

10.	Bolivia	2990	138.7	21548
11.	Uruguay	2828	96.6	29275
12.	South Africa	1170	73.0	16021

Source: FAOSTAT [26].

However, Africa occupies 1-3% of the total world area under soybean production. South Africa leads as the highest producer of soybeans in Africa, followed by Nigeria and Zambia [26]. Its production is constrained by diverse environmental stresses like salinity, drought, high temperature, water logging and biotic factors [33]. Biotic stressors, particularly insect pests, diseases, and weeds, have constantly contributed to severe yield losses thus affecting the quality of soybeans produced [33].

BIOTIC CONSTRAINTS TO COWPEA AND SOYBEAN PRODUCTION

Fungal Diseases

Reports have revealed that more than forty species of fungi cause diseases in cowpeas while soybean is attacked by sixty-six fungi with losses ranging from 20 to 100% [20, 34]. Among these fungal diseases are Fusarium wilt caused by *Fusarium oxysporium*, damping off caused by *Pythium ultimum*, macrophomina blight by *Macrophomina phaseolina*, web blight and root rot by *Thanatephorus cucumeris* (*Rhizoctonia solani*), stem rot by *Phytophthora vignae,* anthracnose diseases and cercospora leaf spot by *Pseudocercospora cruenta* and *Cercospora apii*, respectively [22]. Some of these diseases are also discussed below (Fig. **3**):

Fusarium Wilt

Fusarium wilt is a universal disease that leads to heavy yield losses in crops. Its causal pathogen, *Fusarium* spp., can also induce root rot in various leguminous plants. Factors such as soil nutrient levels, temperature and moisture stress influence the occurrence and epidemic spread of these fungal species [36]. In cowpea, the disease is caused by *Fusarium oxysporum* f. sp. *tracheiphilum* (*Fot*) [35]. This pathogen can cause substantial yield reductions in cowpea production of about 30% to 100%. Currently, only four strains of *F. oxysporum* f. sp. *tracheiphilum* (*Fot*) have been characterized according to their differential interactions on several cowpea genotypes [36]. Although the most widely distributed is *Fot* race 3, race 1 which is known to be predominant in Nigeria [36]. This strain remains very difficult to manage as it commonly appears like diseases caused by other soil-borne pathogens. However, different measures of disease management have been studied which include crop rotation, application of fungicides, and use of biological control agents [36, 37].

In soybeans, *Fot* can also influence massive yield losses. A study revealed that two soybean cultivars susceptible to the pathogen incurred 47.6 and 55.6% yield losses following their test inoculation. In the United States, the estimated losses of soybean yield due to this disease reached 34.3 million bushels/year [38]. Although nineteen different *Fusarium* species have been identified to affect soybean of which *Fusarium oxysporum* and *Fusarium solani* are generally the most devastating in the many countries of the world including the United States [39], the main race that has been commonly known to attack soybean is *Fusarium oxysporum* f. sp. *glycines Fog* [35]. A ubiquitous soil-borne pathogen, *Fot* promotes vascular wilts in young cowpea plants. It has been considered fifth in the top 10 plant pathogens of economic importance [35, 36]. *Fot* displays an asexual pattern of reproduction and can survive extended periods in the soil as chlamydospores when a host is absent.

However, in the presence of a host plant (*e.g.,* cowpea), infection is initiated in the vascular system *via* entry through the roots [35]. Root penetration occurs through natural openings at the intercellular junctions of cortical cells or through wounds [40]. By producing cell wall degrading enzymes, the fungus penetrates the endodermis using the hyphae structure to enter into the xylem vessels [41]. Once inside these vascular channels, it initiates and maintains movement and multiplication simultaneously, hence colonizing the host until the plant begins to show symptoms of witling [42]. Within a short period of time, characteristic symptoms such as leaf epinasty, vascular browning in stems and defoliation are observed in infected cowpea plants [35, 42]. Wilting is induced by a combined activity of the fungus (due to the accumulation of mycelium) and the host defense responses (as gels and gum production) resulting in a blockage of the vascular vessels [35].

Anthracnose Blight

Anthracnose blight is a very destructive disease that occurs in warm and humid regions of the tropics and subtropics. It is caused mainly by *Colletotrichum* sp., pathogens that are transmitted through seeds. It induces severe blight conditions on stems and even pods of economically important legumes which include soybean and cowpea [43]. The disease thrives well and becomes destructive under low temperatures, high humidity and free moisture [44]. Cowpea production is severely constrained by anthracnose disease. *C. destructivum* O' Gara and *C. lindemuthianum* are the major species that devastate the cowpea cropping system. In cowpea fields, yield losses can be up to 35-50%, if disease severity is very high [44]. In Nigeria, 75% reductions in yield have been reported [45]. The disease has also been observed on soybeans with many species of *Colletotrichum* held responsible for the severe damage in crops. A number of species have been reported to attack soybean which include *C. truncatum, C. destructivum, C. liriopes, C. tofieldiae* and *C. verruculosum* [46].

However, *C. truncatum* is considered to be responsible for heavy yield losses in soybean fields [47]. Significant grain yield reductions have been observed and range from 16% to 26% in the United States of America, 30-50% in Thailand, and up to 100% in India [48]. *C. destructivum* and *C. truncatum* are seed-borne and have the ability to over-winter in infected crop debris; hence the difficulty in eradicating them. Both pathogens are polycyclic and can survive for a longer period of time without a host [44, 48]. The anthracnose disease can induce damping-off of soybean seedlings and characteristic symptoms constitute dark depressed and irregular lesions on cotyledons, stems, petioles, and pods. The presence of unique necrotic patterns on the leaf abaxial veins is another diagnostic feature [48]. Specifically on cowpeas, *C. destructivum* sporulates at localized infection foci gradually inducing the formation of small angular brown spots on leaf petioles and veins, all coalescing into brick red to brown discoloration of the entire leaf [44].

Cowpea/ Soybean Rust

Rust diseases are economically important diseases of crops with high devastating ability and can influence massive yield losses if widespread in legume-grown fields. Such diseases are mainly caused by a number of fungal pathogens of which *Phakopsora pachyrhizi* and *Uromyces phaseoli* var. *vignae*, are significant to soybean and cowpea production, respectively. *P. pachyrhizi,* specifically, can parasitize more than 90 legumes causing severe damage [49]. Rust diseases have been reported to reduce yields in soybean cultivations in Brazil by an estimated 2.2 million metric tons [50]. Cowpea rust is considered the most important biotic constraint to the high yield and production of cowpea causing severe loss of 60-80%. The severity of the disease is greater in dry, warm, and highly humid regions of the world. It is caused by the fungal pathogen *Uromyces phaseoli* var. *vignae* [51].

The fungal pathogen infects the leaf surface as air-borne urediniospores which produce toxins such as tentoxin and tabtoxin to inhibit some enzymatic processes that are directly or indirectly required for photosynthesis. Through multiplication and dispersal *via* air currents, several leaves are infected and massive premature defoliation commences causing mortality of young and mature plants [51]. Soybean rust is a foliar disease caused by the fungal pathogen *P. pachyrhizi*, prevalent in tropical and subtropical regions, and becomes severe when it is established under optimum temperature (18-26 °C) with high relative humidity levels (75-80%) [30]. In Asia, yield losses caused by this disease can be as high as

40-80% even with a low disease incidence [52]. In 2012, Brazil lost over 300,000 tonnes of soybeans to the impact of this disease [49].

The disease is airborne as the urediniospore of the fungus is transmitted by prevailing winds and lands on the upper surface of the leaves. Infection is initiated by the urediniospores, and its successive proliferation increases within seven days leading to multiple infected sites [30]. Pathogen spread is favoured by frequent rainfall and heavy dews [50]. The fungus is an obligate biotroph and inhibits the photosynthetic activities of the leaves by forming appressoria (peg-like hyphae) in plant tissues [29]. The uredina ruptures the plant epidermis and causes tissue desiccation and premature defoliation by diminishing stomata regulation of transpiration. Successive replication of symptoms in the soybean plant gradually leads to its rapid death [50]. For rust diseases, different control measures have been investigated such as the use of resistant varieties, crop rotation and fungicide application. However, their efficiency was reduced due to factors like favourable climate for the pathogen, high plant density and the presence of a secondary host [53].

Bacterial Diseases

Many devastating diseases of legumes are caused by bacteria including wilts, soft rots, stem and leaf spots as well as many cankers. The majority of these diseases are mainly due to Gram-negative plant bacterial pathogens including *Pseudomonas* spp., *Ralstonia solanacearum*, *Erwinia* sp. and *Xanthomonas* sp [54] (Fig. **3**). In legumes, bacterial diseases can be classified into two main groups: leaf blights or spots and bacterial wilts [54]. Two bacterial diseases will be discussed further.

Leaf Blight

Leaf blights and spots are mainly caused by two bacterial pathogens, *Pseudomonas syringae* and *Xanthomonas axonopodis* species. A large number of pathovars exist for both species depending on their specific host crop [54]. Bacterial blight caused by *X. axonopodis* pv. *vignicola* (Xav), [Burkholder] Dye and bacterial pustule (*X. axonopodis* pv. *vignae*) are the main bacterial diseases of cowpea [55]. In some areas of West Africa such as Nigeria, yield depression due to *X. axonopodis* may be as high as 71% in pod, 68% in seed, and 53% in fodder with severe grain yield loss of more than 64% were reported. This indicates that the causative pathogens thrive well under warm humid climatic conditions which is unique to tropical regions of the world [56]. Obvious symptoms of infection by both pathogen include large, irregular foliar lesions with yellow margins; stem cankers, flower and pod abortion, leaf defoliation as well as pre-emergent and post-emergent seedling mortality [57].

Fig. (3). Examples of fungal and bacterial diseases affecting soybean and cowpea. **From top right to left:** Fusarium wilt on cowpea plants, Anthracnose blight on cowpea and Soybean stems severely damaged by *Fusarium oxysporum.* **From middle left to right:** Stems of cowpea affected by *F. oxysporum,* Leaf rust of cowpea and Symptoms of rust on soybean. **From bottom right to left:** Damping off disease on cowpea roots caused by *Ralstonia solani, Xanthomonas axonopodis* affecting organ parts of cowpea and *Pseudomonas* blight on soybean leaf [10, 16, 45, 53].

Symptoms begin with small water-soaked spots which gradually coalesce into necrotic lesions surrounded by yellow haloes. The stem is also invaded by the pathogen, producing cracks with brown stripes, swollen canker, dark green water-soaked patches on pods and finally discoloration in seeds [58]. Soybean bacterial blight is caused by *Pseudomonas syringae* pv. *Glycinea,* known resulting in seed losses of up to 28%. Under extreme conditions, the yield loss potential for this disease ranges from 4% to a maximum of 40% [59]. Young leaves have been observed to be more susceptible to bacterial infection, hence characteristic lesions are small yellow to brown spots on leaf surfaces. Gradually the lesion dries out, turning reddish brown to black with a surrounding yellowish-green halo [59]. Enlargement of the small lesions could occur to produce larger irregular necrotic areas. *P. syringae* can also infect stem petioles and pods; hence its seed-borne transmission, although its main means of dispersal is *via* wind or splashing water droplets [59, 60].

Viral Diseases

Viruses are obligate subcellular parasites that rapidly multiply in living cells and frequently induce different disease conditions. They are only transmitted and dispersed through insect vectors or infected planting materials (seeds). For legumes, the most important vectors of viruses are aphids and whiteflies [54]. Every year throughout the world, they cause economic losses estimated at up to US $ 60 billion [61]. Several viruses infect cowpeas, which include cowpea aphid-borne mosaic virus (CABMV), cowpea mosaic virus (CPMV), cowpea mild mottle virus (CPMMV) and cucumber mosaic virus (CMV) [55]. Cowpea aphid-borne mosaic (genus *Potyvirus*, family *Potyviridae*) is a cosmopolitan, economically significant seed-borne virus of cowpea. It can cause yield loss of 13-87% under field conditions depending upon virus strain, crop susceptibility and environmental conditions [62]. As the name implies, it is primarily transmitted, in a non-persistent manner, through aphids (*Aphis craccivora*). The viral infection induces the formation of mosaic patterns with different intensities on the leaves.

Other characteristic symptoms include the appearance of yellowish ring-shaped spots, mottling, roughness, and blistering, shortening of internodes and deformation of pods [63]. Also, the cowpea mild mottle virus (CPMMV, genus *Carlavirus*, family *Betaflexivirdae*) is economically significant to cowpea production worldwide. It is transmitted by seeds of infected plants and several weeds. It is also disseminated by whiteflies (*Bemisia tabaci*, Gennadius) in a semi-persistent manner [64]. Symptoms of its infection on cowpeas include the appearance of necrotic lesions on the primary leaves, severe systemic chlorosis and necrosis on trifoliate leaves [64]. However, the bean pod mottle virus (BPMV genus *Comovirus*, family *Comoviridae*) and soybean mosaic virus (SMV genus *Potyvirus*, family *Potyviridae*) are among the most devastating viral pathogens in soybean [54].

The BPMV is a single-stranded RNA virus affecting soybean production globally and it is transmitted by the leaf beetle (*Cerotorma trificurta* Foster). Other potential sources of its inoculum can include seed-to-seedling transmission and alternative leguminous weed hosts. It is reported to reduce yield by up to 40%, having subjected the soybean crop to soil water stress as well. Infection at vegetative soybean stage one (V1) can result in yield losses of about 52.6% [65]. The infection is initiated when inoculative beetles deposit the active virus in regurgitation on the surface of the wounded leaf during feeding. Afterward, through the leaf tissues and *via* the xylem, the virus particles are translocated to unaffected plant cells [66]. Characteristic symptoms of BPMV infection include severe systemic mottling with a puckering of leaves and mottling of pods and seed coats [67].

SMV, on the other hand, is also a single-stranded positive-sense RNA virus that is transmitted by aphids (*Aphis glycines*) and *via* infected seeds. Infection commences through the feeding activity of the aphids on infected plant tissues [68]. Viruliferous aphids introduce the virus into healthy plant parts through natural openings or physical wounds. On entry, the virus' genomic RNA is released and translated. Following the translation of viral proteins, virus particles reassemble, and new virus progeny move into neighbouring cells [69]. The appearance of yellow vein clearing on new trifoliate leaves is observed as a symptom followed by a downward and upward curving of leaf margins and tips respectively. Leaves begin to appear brittle before maturity, mottling sets in during cold weather conditions, and lesions emerge on leaf surfaces [68] (Fig. 4).

Fig. (4). Some disease symptoms are caused by virus and plant parasitic nematode. **From top left to right:** Mosaic patterns on cowpea leaves caused by cowpea mosaic virus, Soybean leaves infected with soybean mosaic virus and Mild mottle virus disease on cowpea leaves. **From bottom right to left:** Soybean cyst nematode symptoms on soybean roots, Root galls on cowpea roots caused by *Meloidogyne incognita* and Soybean pods severely damaged by bean pod mottle virus [63, 64, 80, 81].

Weeds

Parasitic weeds are plants known to attack other plants by making connections and deriving part or all of their food from the host. They achieve this using specialized structures that allow for attachment to the roots or shoots of the host plant [70]. Compared to other major pests of agricultural crops, weeds alone have caused severe yield losses ranging from as low as 10% to as high as 98% of total crop failure in the dry land regions. The production potential of many grain legumes has been hampered by the activity of various parasitic weeds [70]. The most economically important parasitic weeds are the broomrapes (*Orobanche*), *Striga,* and dodder. Amongst these, *Striga* can be very devastating. *Striga* (witchweed) is mainly present in the dry savanna area of West and Central Africa, hence it is a major pest to cowpeas production than soybean [55]. Although there are about 28 species in this genus, *Striga gesneriodes* (wild) Vatke attacks cowpeas and can cause grain yield loss of up to 100% in cases of high population density of the weed. *S. gesnerioides*, having no functional leaves, draws a wider range of nutrients from its host using a specialized piercing and sucking structure called the haustoria. The increasing incidence of *Striga* has been attributed to poor soil fertility and structure, as well as moisture stress [70]. Its parasitism leads to the weakening of the cowpea plants, wounding of its outer root tissues, and consequently the absorption of its supply of moisture, photosynthates and minerals. Few reports and scanty scientific information suggest that soybean seems not to be much challenged by parasitic weeds (Fig. **5**).

Insect Pests of Cowpea and Soybean

Insects are important pests of legumes and can cause tremendous damage because of their high fertility and short generation times. Such insects can be categorized into chewing, piercing, sucking and boring based on mouth parts or pre-harvest and post-harvest based on crop developmental stage. Examples of chewing insect pests include bruchid weevil (*Callosobruchus* sp.) and pod borers (*Helicoverpa* sp. and *Nezara* sp.) while for sucking insects we have aphids (*Aphis* sp.), and whiteflies (*Bemisia* sp.) [54]. A few will be discussed with respect to their significant effect on cowpea and soybean production. The field insect pests of major economic importance within the cowpea cropping agro-ecologies worldwide are *A. craccivora* (cowpea aphid), *Megalurothrips sjostedti* (flower bud thrips), *Maruca vitrata* (spotted pod borer) and *C. maculatus* [71]. *A. craccivora* Koch (Homoptera: Aphididae) is a key foliage pest whose adults and nymphs feed on the under-surface of young leaves, stem tissues, petioles, flowers and pods of mature plants by sucking fluid. Severe infestation leads to stunting, leaf defoliation and death of seedlings [72].

Fig. (5). Insect pests and parasitic weeds affecting soybean and cowpea. **From top left to right:** Stink bug on sowpea leaf surface (*Nezara viridula*), Cowpea leaves severely infested with whiteflies (*Bemisia tabaci*) and Larvae of pod borer (*Helicoverpa armigera*) on soybean. **From bottom right to left:** Striga parasitic weed competing with cowpea plants, Cowpea weevils (*Callosobruchus maculatus*) feeding on cowpea grains and Pod borer (*Maruca vitrata*) emerging from cowpea seed after inducing severe damage [73, 74, 76].

Besides, they are responsible for the transmission of the cowpea aphid-borne mosaic virus and infestation by these field pests resulting in yield losses of about 50% [72]. *M. sjostedti* Tryborn (Thysanoptera: Thripidae) is the first major insect pest of cowpea at the reproductive stage. Leaf and flower buds are attacked by the insect resulting in browning, distortion, abscission and abortion of flower buds. Infestations are severe in dry seasons as the population density of the insect could increase. A yield loss of 100% can be incurred by the devastating activity of *M. sjostedti* [55, 72]. The legume pod borer, *M. vitrata* Fabricius (Lepidoptera: Crambidae) is a pantropical and post-flowering cowpea pest. It is capable of causing extensive grain yield reduction (20%–80%) if not controlled. Larvae feed on tender parts of the stem, flower buds, peduncles and pods. By forming webs around these organs, flower abortion is provoked [72, 73]. In grain storage, cowpea is severely attacked by the larvae of bruchids weevils (*C. maculatus*) which causes extensive damage by feeding and developing inside which feed and develop inside the seeds, with adult insects boring holes through which the new insects emerge.

Yield losses in terms of grain quality and quantity can amount to over 90% when there's no pre-storage treatment with insecticide [72, 74]. In soybeans, *Aphis glycines* Matsumura (Hemiptera: Aphididae) has been known to cause major losses. Soybean aphids feed on the above-ground plant parts (leaves and stems); hence reducing photosynthate production and yield [29]. Symptoms displayed by infected plants are similar to those of the cowpea aphid. They also indirectly transmit different pathogens (especially viruses), and yield reductions of about 40% as a result of *A. glycines* have been reported [75]. Soybean is also affected by stink bugs, *Piezodorus guildinii* (Hemiptera: Pentatomidae), which are shield-shaped insects that feed on the pods of soybean [29]. Also called red-banded stink bug, it has a broad geographical range. The adult and juvenile stages feed directly on seeds which leads to pod deformation. Their sucking activity creates punctures in pods that favour the occurrence of fungi such as *Nematospora coryli*, the causal pathogen of yeast spot [76].

Nematode Pests of Cowpea and Soybean

Plant parasitic nematodes (PPNs) are obligate parasites known to affect the yield and productivity of some legumes such as groundnut, soybean and cowpea [77]. The impact of PPNs is felt more due to their ability to expose the infected plant to secondary infections. A number of soil-borne pathogens such as *Fusarium* sp. and *Verticillium* sp. obtain entry into plants through wounds created on their roots by the parasitic activity of PPNs. With such interaction, they can induce 7-15% yield losses in agricultural crops [78]. Two PPNs will be discussed based on their economic importance to the production of soybean and cowpea: the soybean cyst nematode and the root-knot nematode. The soybean cyst nematode, *Heterodera glycines*, (SCN) is an obligate, ubiquitous, and sedentary endoparasite inhabiting the soil. It is considered a serious nematode pest in many soybean-producing countries known to cause sudden death syndrome (SDS). It has been reported to cause approximately 36% of yield losses in total soybean production in the United States of America from 1996 to 2014 [79]. Economic losses of over $120 million have also been reported in the USA [80]. Infection is initiated by the second juvenile stage of SCN which hatches from the egg in the soil and gets attracted to the roots of soybean plants. Using a hollow mouth spear (stylet), it pierces through the cell wall of the root cells while secreting digestive enzymes to aid intracellular migration through the cortex toward the vascular cylinder. It then selects a single cell near the vascular tissue as a permanent feeding leading to the formation of cysts [80]. In infested soybean fields, foliar symptoms appear, reproductive stage is chlorotic mottling and necrosis along leaf veins which gradually evolves to premature defoliation, stunting, and early maturity [81]. Although these visible above-ground symptoms may be confused with other stresses, the observation of cysts on soybean roots confirms the presence of SCN.

Yield losses are more severe when *Fusarium virguliforme* infect soybean seedling roots shortly after germination causing root rots and reduced root volume [81].

The root-knot nematode *Meloidogyne* sp., (RKN) is considered one of the most devastating nematode species, having a wide host range which includes cereals, leguminous crops, vegetables and trees [82]. Among the root-knot nematode species, *M. incognita* (Kofoid & White) Chitwood, *M. arenaria* (Neal), *M. javanica* (Treub) Chitwood, and *M. hapla* (Chitwood) are the common nematodes causing 95% of the incidence of RKN parasitism on important crop plants worldwide. *M. incognita* is the most common and destructive to soybeans and other important crops, including tobacco, peanuts, and cowpeas [79]. The mechanism of infection by RKNs is similar to that of SCNs. The second infective juvenile stage (J2) hatch from their respective eggs, and penetrate the roots directly behind the root cap using a combination of physical damage through thrusting of the stylet and breakdown of the cell wall by cellulolytic and pectolytic enzymes.

In contrast to SCNs, the J2 migrate intercellularly within the roots, find and initiate a permanent feeding within the vascular cylinder which leads to the formation of giant cells (outwardly seen as galls). These galls (in larger numbers) serve as specialized sinks drawing nutrients from the plant [83]. The appearance of multiple galls on the roots of soybeans and cowpeas is the main characteristic symptom of infection. Above-ground symptoms include chlorosis on leaves, stunted growth, root galling, and excessive branching of the root [84]. In the United States of America alone, yield losses of soybeans due to RKNs from 1974 to 2007 were up to approximately 2.1 million tonnes with an annual economic loss of almost $30 million experienced in soybeans [79]. However, in cowpeas, yield losses of 73-100% have been reported in Northern Ghana as a result of infection by RKNs. Heavy infestation of cowpea by *Meloidogyne incognita* leads to early senescence of the crop. In high populations, damage by RKNs can also induce yield losses of more than 90% [84].

CONCLUSION

Legumes are very important agricultural crops that are critical to ensuring food security in the world, especially in developing countries. Diverse environmental stresses are known to limit optimal production of economically significant legumes. Biotic stressors such as insects, nematodes, fungi, bacteria, virus and parasitic weeds are sorely responsible for the major losses incurred by these crops. As such, grain legumes like cowpea and soybean will continue to face the impact of these biotic constraints, which have influenced massive yield losses and as a result threaten food security across the globe.

LIST OF ABBREVIATIONS

CABMV Cowpea aphid-borne mosaic virus

CPMV Cowpea mosaic virus

CPMMV Cowpea mild mottle virus

CMV Cucumber mosaic virus

FAOSTAT Food and Agriculture Organisation of the United Nations Statistics

PPNs Plant parasitic nematodes

RKN Root-knot nematode

SCN Soybean cyst nematode

SMV Soybean mosaic virus

SDS Sudden death syndrome

ss-RNA single-stranded Ribonucleic acid

USDA United States Department of Agriculture

CONSENT FOR PUBLICATION

Not applicable.

CONFLICT OF INTEREST

The author declares no conflict of interest, financial or otherwise.

ACKNOWLEDGEMENT

Declared none.

REFERENCES

[1]   Rodríguez-Sifuentes, L.; Marszalek, J.E.; Chuck-Hernández, C.; Serna-Saldívar, S.O. Legumes protease inhibitors as biopesticides and their defense mechanisms against biotic factors. *Int. J. Mol. Sci.,* **2020**, *21*(9), 3322.
       [http://dx.doi.org/10.3390/ijms21093322] [PMID: 32397104]

[2]   Singh, B.; Singh, J.P.; Shevkani, K.; Singh, N.; Kaur, A. Bioactive constituents in pulses and their health benefits. *J. Food Sci. Technol.,* **2017**, *54*(4), 858-870.
       [http://dx.doi.org/10.1007/s13197-016-2391-9] [PMID: 28303037]

[3]   Rebello, C.J.; Greenway, F.L.; Finley, J.W. A review of the nutritional value of legumes and their effects on obesity and its related co-morbidities. *Obes. Rev.,* **2014**, *15*(5), 392-407.
       [http://dx.doi.org/10.1111/obr.12144] [PMID: 24433379]

[4]   Kebede, E. Grain legumes production and productivity in Ethiopian smallholder agricultural system, contribution to livelihoods and the way forward. In: *Cogent Food Agric*; Yildiz, F., Ed.; , **2020**; 6, p. (1)1722353.
       [http://dx.doi.org/10.1080/23311932.2020.1722353]

[5]   Watson, C.A.; Reckling, M.; Preissel, S.; Bachinger, J.; Bergkvist, G.; Kuhlman, T. Grain legume production and use in european agricultural systems. In: *Advances in Agronomy*; Elsevier, **2017**; pp.

235-303. Available from: https://linkinghub.elsevier.com/retrieve/pii/S0065211317300202

[6]   Teshome, B. The Traditional Practice of Farmers' Legume-Cereal Cropping System and the Role of Microbes for Soil Fertility Improvement in North Shoa, Ethiopia. *Agric Res Technol Open Access J,* **2018,** *13*(4) Available from: https://juniperpublishers.com/artoaj/ARTOAJ.MS.ID.555891.php (cited 2021 Sep 21).

[7]   Rubiales, D.; Mikic, A. Introduction: Legumes in sustainable agriculture. *Crit. Rev. Plant Sci.,* **2015,** *34*(1-3), 2-3.
        [http://dx.doi.org/10.1080/07352689.2014.897896]

[8]   Araújo, S.S.; Beebe, S.; Crespi, M.; Delbreil, B.; González, E.M.; Gruber, V.; Lejeune-Henaut, I.; Link, W.; Monteros, M.J.; Prats, E.; Rao, I.; Vadez, V.; Patto, M.C.V. Abiotic stress responses in legumes: Strategies used to cope with environmental challenges. *Crit. Rev. Plant Sci.,* **2015,** *34*(1-3), 237-280.
        [http://dx.doi.org/10.1080/07352689.2014.898450]

[9]   Suzuki, N.; Rivero, R.M.; Shulaev, V.; Blumwald, E.; Mittler, R. Abiotic and biotic stress combinations. *New Phytol.,* **2014,** *203*(1), 32-43.
        [http://dx.doi.org/10.1111/nph.12797] [PMID: 24720847]

[10]   Gimenez, E.; Salinas, M.; Manzano-Agugliaro, F. Worldwide research on plant defense against biotic stresses as improvement for sustainable agriculture. *Sustainability,* **2018,** *10*(2), 391.
        [http://dx.doi.org/10.3390/su10020391]

[11]   Farooq, M.; Hussain, M.; Usman, M.; Farooq, S.; Alghamdi, S.S.; Siddique, K.H.M. Impact of abiotic stresses on grain composition and quality in food legumes. *J. Agric. Food Chem.,* **2018,** *66*(34), 8887-8897.
        [http://dx.doi.org/10.1021/acs.jafc.8b02924] [PMID: 30075073]

[12]   Kasper, S.; Christoffersen, B.; Soti, P.; Racelis, A. Abiotic and biotic limitations to nodulation by leguminous cover crops in south texas. *Agriculture,* **2019,** *9*(10), 209.
        [http://dx.doi.org/10.3390/agriculture9100209]

[13]   Benjamin, J.; Adejumo, S.A.; Claudius-Cole, A. Maize response to sole and combined effects of nitrogen and nematode stresses. *Adv.J. Grad. Res.,* **2020,** *9*(1), 71-80.
        [http://dx.doi.org/10.21467/ajgr.9.1.71-80]

[14]   Nadeem, M.; Li, J.; Yahya, M.; Sher, A.; Ma, C.; Wang, X.; Qiu, L. Research progress and perspective on drought stress in legumes: A review. *Int. J. Mol. Sci.,* **2019,** *20*(10), 2541.
        [http://dx.doi.org/10.3390/ijms20102541] [PMID: 31126133]

[15]   Gull, A.; Ahmad Lone, A.; Ul Islam Wani, N. Biotic and Abiotic Stresses in Plants. In: *Abiotic and Biotic Stress in Plants*; Bosco de Oliveira, A., Ed.; IntechOpen, **2019**. Available from: https://www.intechopen.com/books/abiotic-and-biotic-stress-in-plants/biotic-and--biotic-stresses-in-plants (cited 2021 Sep 20).
        [http://dx.doi.org/10.5772/intechopen.85832]

[16]   Pande, S; Sharma, SB; Ramakrishna, A Biotic stresses affecting legumes production in the Indo-Gangetic Plain. In: Legumes in rice and wheat cropping systems of the Indo-Gangetic Plain - constraints and opportunities. In: *International Crops Research Institute for the Semi-Arid Tropics*; Patancheru: Andhra Pradesh, India, **2000**; pp. 129-155.

[17]   Li, H.; Rodda, M.; Gnanasambandam, A.; Aftab, M.; Redden, R.; Hobson, K.; Rosewarne, G.; Materne, M.; Kaur, S.; Slater, A.T. Breeding for biotic stress resistance in chickpea: Progress and prospects. *Euphytica,* **2015,** *204*(2), 257-288.
        [http://dx.doi.org/10.1007/s10681-015-1462-8]

[18]   Rasool, S.; Abdel Latef, A.; Ahmad, P. Chickpea. In: *Legumes under Environmental Stress*; Wiley Online Library, **2015**; pp. 67-79.

[19]   Huynh, B-L; Ehlers, JD; Close, TJ; Cisse, N; Drabo, I; Lucas, MR Enabling tools for modern breeding

of cowpea for biotic stress resistance. *Translational Genomics for Crop Breeding,* **2015**, , 183-199.

[20] Horn, L.N.; Shimelis, H. Production constraints and breeding approaches for cowpea improvement for drought prone agro-ecologies in Sub-Saharan Africa. *Ann. Agric. Sci.,* **2020**, *65*(1), 83-91.
[http://dx.doi.org/10.1016/j.aoas.2020.03.002]

[21] Ibro, G.; Sorgho, M.C.; Idris, A.A.; Moussa, B.; Baributsa, D.; Lowenberg-DeBoer, J. Adoption of cowpea hermetic storage by women in Nigeria, Niger and Burkina Faso. *J. Stored Prod. Res.,* **2014**, *58*, 87-96.
[http://dx.doi.org/10.1016/j.jspr.2014.02.007]

[22] Badiane, F.A.; Diouf, M.; Diouf, D. Cowpea. In: *Broadening the Genetic Base of Grain Legumes*; Singh, M.; Bisht, I.S.; Dutta, M., Eds.; Springer India: New Delhi, **2014**; pp. 95-114.
[http://dx.doi.org/10.1007/978-81-322-2023-7_5]

[23] Odireleng, O.M.; Chiyapo, G.; Joshuah, M.; Stephen, M.C. Phenotypic variation in cowpea *(Vigna unguiculata [L.] Walp.)* germplasm collection from Botswana. *Int. J. Biodivers. Conserv.,* **2016**, *8*(7), 153-163.
[http://dx.doi.org/10.5897/IJBC2016.0949]

[24] Daryanto, S.; Wang, L.; Jacinthe, P-A. Global synthesis of drought effects on food legume production. *PLoS One,* **2015**, *10*(6), e0127401.
[http://dx.doi.org/10.1371/journal.pone.0127401]

[25] Horn, L.; Shimelis, H.; Laing, M. Participatory appraisal of production constraints, preferred traits and farming system of cowpea in the northern Namibia: Implications for breeding. *Legume Res - Int J,* **2015**, *38*(5). Available from: http://arccjournals.com/index.php?option=com_journals&view= article&id=6275&Itemid=641 (cited 2021 Sep 22).
[http://dx.doi.org/10.18805/lr.v38i5.5952]

[26] Retrieved, F.A.O.S.T.A.T. Countries- Select All; Regions-World + (Total); Elements- Production Quantity; Items- Soybeans and Cowpea. *Year,* **2021**, *2019*, 2019.

[27] USDA Food and Nutrient Database for Dietary Studies (FNDDS). Cowpea (mature seeds, raw) nutritional composition. FoodData Central. **2018**. Available from: https://fdc.nal.usda.gov/fdc-app.html#/food-details/175208/nutrients

[28] International Institute of Tropical Agriculture (IITA). *Nigeria, Kamara AY, Omoigui LO, International Institute of Tropical Agriculture (IITA), Nigeria, Kamai N, International Institute of Tropical Agriculture (IITA), Nigeria, et al. Improving cultivation of cowpea in West Africa. In: Burleigh Dodds Series in Agricultural Science*; Burleigh Dodds Science Publishing, **2018**, pp. 235-252. Available from: https://shop.bdspublishing.com/store/bds/detail/product/3-190-9781838794170 (cited 2021 Aug 29).

[29] Hartman, G.; Pawlowski, M.; Herman, T.; Eastburn, D. Organically grown soybean production in the usa: Constraints and management of pathogens and insect pests. *Agronomy,* **2016**, *6*(1), 16.
[http://dx.doi.org/10.3390/agronomy6010016]

[30] Murithi, H.M.; Beed, F.; Tukamuhabwa, P.; Thomma, B.P.H.J.; Joosten, M.H.A.J. Soybean production in eastern and southern Africa and threat of yield loss due to soybean rust caused by *Phakopsora pachyrhizi. Plant Pathol.,* **2016**, *65*(2), 176-188.
[http://dx.doi.org/10.1111/ppa.12457]

[31] Agarwal, D.K.; Billore, S.D.; Sharma, A.N.; Dupare, B.U.; Srivastava, S.K. Soybean: Introduction, improvement, and utilization in india—problems and prospects. *Agric. Res.,* **2013**, *2*(4), 293-300.
[http://dx.doi.org/10.1007/s40003-013-0088-0]

[32] Sinclair, T.R.; Marrou, H.; Soltani, A.; Vadez, V.; Chandolu, K.C. Soybean production potential in Africa. *Glob. Food Secur.,* **2014**, *3*(1), 31-40.
[http://dx.doi.org/10.1016/j.gfs.2013.12.001]

[33] Hasanuzzaman, M.; Nahar, K.; Rahman, A.; Mahmud, J.A.; Hossain, M.S.; Fujita, M. Soybean production and environmental stresses. In: *Environmental Stresses in Soybean Production*; Elsevier, **2016**;   pp.   61-102.   Available   from: https://linkinghub.elsevier.com/retrieve/pii/B9780128015353000048
[http://dx.doi.org/10.1016/B978-0-12-801535-3.00004-8]

[34] Rahman, M.; Rubayet, MdT.; Bhuiyan, K. Integrated management of fusarium root rot and wilt disease of soybean caused by Fusarium oxysporum. *Int J Biosci IJB.,* **2020**, *17*, 83-96.

[35] Sampaio, A.M.; Araújo, S.S.; Rubiales, D.; Vaz Patto, M.C. Fusarium wilt management in legume crops. *Agronomy,* **2020**, *10*(8), 1073.
[http://dx.doi.org/10.3390/agronomy10081073]

[36] Omoigui, L.; Danmaigona, C.; Kamara, A.; Ekefan, E.; Timko, M. Genetic analysis of Fusarium wilt resistance in cowpea *(V igna unguiculata Walp.). Plant Breed.,* **2018**, 137.

[37] Zhang, C.; Wang, W.; Xue, M.; Liu, Z.; Zhang, Q.; Hou, J.; Xing, M.; Wang, R.; Liu, T. The combination of a biocontrol agent *Trichoderma asperellum* SC012 and hymexazol reduces the effective fungicide dose to control fusarium wilt in cowpea. *J. Fungi,* **2021**, *7*(9), 685.
[http://dx.doi.org/10.3390/jof7090685] [PMID: 34575723]

[38] Arias, M.M.D.; Leandro, L.F.; Munkvold, G.P. Aggressiveness of *Fusarium* species and impact of root infection on growth and yield of soybeans. *Phytopathology,* **2013**, *103*(8), 822-832.
[http://dx.doi.org/10.1094/PHYTO-08-12-0207-R] [PMID: 23514263]

[39] Arias, M.M.D.; Munkvold, G.P.; Ellis, M.L.; Leandro, L.F.S. Distribution and frequency of *Fusarium* species associated with soybean roots in iowa. *Plant Dis.,* **2013**, *97*(12), 1557-1562.
[http://dx.doi.org/10.1094/PDIS-11-12-1059-RE] [PMID: 30716864]

[40] Michielse, C.B.; Rep, M. Pathogen profile update: *Fusarium oxysporum. Mol. Plant Pathol.,* **2009**, *10*(3), 311-324.
[http://dx.doi.org/10.1111/j.1364-3703.2009.00538.x] [PMID: 19400835]

[41] Pérez-Nadales, E.; Di Pietro, A. The membrane mucin Msb2 regulates invasive growth and plant infection in *Fusarium oxysporum. Plant Cell,* **2011**, *23*(3), 1171-1185.
[http://dx.doi.org/10.1105/tpc.110.075093] [PMID: 21441438]

[42] Perez-Nadales, E.; Almeida Nogueira, M.F.; Baldin, C.; Castanheira, S.; El Ghalid, M.; Grund, E.; Lengeler, K.; Marchegiani, E.; Mehrotra, P.V.; Moretti, M.; Naik, V.; Oses-Ruiz, M.; Oskarsson, T.; Schäfer, K.; Wasserstrom, L.; Brakhage, A.A.; Gow, N.A.R.; Kahmann, R.; Lebrun, M.H.; Perez-Martin, J.; Di Pietro, A.; Talbot, N.J.; Toquin, V.; Walther, A.; Wendland, J. Fungal model systems and the elucidation of pathogenicity determinants. *Fungal Genet. Biol.,* **2014**, *70*(100), 42-67.
[http://dx.doi.org/10.1016/j.fgb.2014.06.011] [PMID: 25011008]

[43] Subedi, S.; Gharti, D.B.; Neupane, S.; Ghimire, T. Management of anthracnose in soybean using fungicide. *Journal of Nep. Agricul. Res. Coun.,* **2016**, *1*, 29-32.
[http://dx.doi.org/10.3126/jnarc.v1i0.15721]

[44] Obi, V.I.; Barriuso-Vargas, J.J. Situation of biofungicides reconnaissance, a case of anthracnose disease of cowpea. *Am. J. Plant Sci.,* **2014**, *5*(9), 1202-1211.
[http://dx.doi.org/10.4236/ajps.2014.59133]

[45] Falade, M.; Enikuomehin, O.; Borisade, O.; Aluko, M. Control of Cowpea *(Vigna unguiculata L. Walp)* diseases with intercropping of maize *(Zea mays L)* and spray of plant extracts. *J. Adv. Microbiol.,* **2018**, *7*(4), 1-10.
[http://dx.doi.org/10.9734/JAMB/2017/38156]

[46] Yang, H.C.; Haudenshield, J.S.; Hartman, G.L. Colletotrichum incanum sp. nov., a curved-conidial species causing soybean anthracnose in USA. *Mycologia,* **2014**, *106*(1), 32-42.
[http://dx.doi.org/10.3852/13-013] [PMID: 24603833]

[47]   Rogério, F.; Gladieux, P.; Massola, N.S., Jr; Ciampi-Guillardi, M. Multiple introductions without admixture of *Colletotrichum truncatum* associated with soybean anthracnose in brazil. *Phytopathology,* **2019**, *109*(4), 681-689.
[http://dx.doi.org/10.1094/PHYTO-08-18-0321-R] [PMID: 30451637]

[48]   Dias, M.D.; Dias-Neto, J.J.; Santos, M.D.M.; Formento, A.N.; Bizerra, L.V.A.S.; Fonseca, M.E.N.; Boiteux, L.S.; Café-Filho, A.C. Current status of soybean anthracnose associated with *Colletotrichum truncatum* in Brazil and Argentina. *Plants,* **2019**, *8*(11), 459.
[http://dx.doi.org/10.3390/plants8110459] [PMID: 31671821]

[49]   Godoy, C.V.; Seixas, C.D.S.; Soares, R.M.; Marcelino-Guimarães, F.C.; Meyer, M.C.; Costamilan, L.M. Asian soybean rust in Brazil: Past, present, and future. *Pesqui. Agropecu. Bras.,* **2016**, *51*(5), 407-421.
[http://dx.doi.org/10.1590/S0100-204X2016000500002]

[50]   Sikora, E.J.; Allen, T.W.; Wise, K.A.; Bergstrom, G.; Bradley, C.A.; Bond, J.; Brown-Rytlewski, D.; Chilvers, M.; Damicone, J.; DeWolf, E.; Dorrance, A.; Dufault, N.; Esker, P.; Faske, T.R.; Giesler, L.; Goldberg, N.; Golod, J.; Gómez, I.R.G.; Grau, C.; Grybauskas, A.; Franc, G.; Hammerschmidt, R.; Hartman, G.L.; Henn, R.A.; Hershman, D.; Hollier, C.; Isakeit, T.; Isard, S.; Jacobsen, B.; Jardine, D.; Kemerait, R.; Koenning, S.; Langham, M.; Malvick, D.; Markell, S.; Marois, J.J.; Monfort, S.; Mueller, D.; Mueller, J.; Mulrooney, R.; Newman, M.; Osborne, L.; Padgett, G.B.; Ruden, B.E.; Rupe, J.; Schneider, R.; Schwartz, H.; Shaner, G.; Singh, S.; Stromberg, E.; Sweets, L.; Tenuta, A.; Vaiciunas, S.; Yang, X.B.; Young-Kelly, H.; Zidek, J. A coordinated effort to manage soybean rust in north america: A success story in soybean disease monitoring. *Plant Dis.,* **2014**, *98*(7), 864-875.
[http://dx.doi.org/10.1094/PDIS-02-14-0121-FE] [PMID: 30708845]

[51]   Tagoe, SMA; Mensah, TA; Asare, AT Effect of Rust *(Uromyces Phaseoli Var. Vignae)* infection on photosynthetic efficiency, growth and yield potentials of cowpea *(Vigna Unguiculata L. Walp)* in an Open Field System. *Glob J Sci Front Res.,* **2020**, 49-61.

[52]   Kawashima, C.G.; Guimarães, G.A.; Nogueira, S.R.; MacLean, D.; Cook, D.R.; Steuernagel, B.; Baek, J.; Bouyioukos, C.; Melo, B.V.A.; Tristão, G.; de Oliveira, J.C.; Rauscher, G.; Mittal, S.; Panichelli, L.; Bacot, K.; Johnson, E.; Iyer, G.; Tabor, G.; Wulff, B.B.H.; Ward, E.; Rairdan, G.J.; Broglie, K.E.; Wu, G.; van Esse, H.P.; Jones, J.D.G.; Brommonschenkel, S.H. A pigeonpea gene confers resistance to Asian soybean rust in soybean. *Nat. Biotechnol.,* **2016**, *34*(6), 661-665.
[http://dx.doi.org/10.1038/nbt.3554] [PMID: 27111723]

[53]   Echeveste da Rosa, C.R. Asian soybean rust resistance: An overview. *J. Plant Pathol. Microbiol.,* **2015**, *6*(9). Available from: https://www.omicsonline.org/open-access/asian-soybean-rust-resist-nce-an-overview-2157-7471-1000307.php?aid=62340
[http://dx.doi.org/10.4172/2157-7471.1000307]

[54]   Rubiales, D.; Fondevilla, S.; Chen, W.; Gentzbittel, L.; Higgins, T.J.V.; Castillejo, M.A.; Singh, K.B.; Rispail, N. Achievements and challenges in legume breeding for pest and disease resistance. *Crit. Rev. Plant Sci.,* **2015**, *34*(1-3), 195-236.
[http://dx.doi.org/10.1080/07352689.2014.898445]

[55]   Boukar, O.; Fatokun, C.A.; Huynh, B.L.; Roberts, P.A.; Close, T.J. Genomic tools in cowpea breeding programs: Status and perspectives. *Front. Plant Sci.,* **2016**, *7*, 757. Available from: http://journal.frontiersin.org/Article/10.3389/fpls.2016.00757/abstract
[http://dx.doi.org/10.3389/fpls.2016.00757] [PMID: 27375632]

[56]   Okechukwu, RU; Ekpo, EJA; Okechukwu, OC Seed to plant transmission of *Xanthomonas campestris* pv. vignicola isolates in cowpea. *Afri. J. Agricul. Res.,* **2010**, *5*(6)

[57]   Amodu, US; Shenge, K; Akpa, DA; Agbenin, ON Population Dynamics of *Xanthomonas axonopodis* pv. vignicola (Burkholder) Dye in/on growing Cowpea Plant. *Int. J. Adv. Acad. Res. Sci. Technol.Eng.,* **2017**, *3*(6), 12.

[58]   Claudius-Cole, A.O.; Ekpo, E.J.A.; Schilder, A.M.C. Evaluation of detection methods for cowpea bacterial blight caused by *Xanthomonas axonopodis* pv vignicola In Nigeria. *Trop. Agric. Res. Ext.,* **2016**, *17*(2), 77.
[http://dx.doi.org/10.4038/tare.v17i2.5310]

[59]   Jagtap, GP; Dhopte, SB; Dey, U Bio-efficacy of different antibacterial antibiotic, plant extracts and bioagents against bacterial blight of soybean caused by *Pseudomonas syringae* pv. glycinea. *Biology,* **2012**.

[60]   Mamgain, S.; Dhiman, S.; Pathak, R.K.; Baunthiyal, M. *In silico* identification of agriculturally important molecule(s) for defense induction against bacterial blight disease in Soybean (Glycine max). *Plant Omics,* **2018**, *11*(2), 98-105.
[http://dx.doi.org/10.21475/poj.11.02.18.1238]

[61]   Zanardo, L.G.; Carvalho, C.M. Cowpea mild mottle virus *(Carlavirus, Betaflexiviridae)*: A review. *Trop. Plant Pathol.,* **2017**, *42*(6), 417-430.
[http://dx.doi.org/10.1007/s40858-017-0168-y]

[62]   Bashir, M.; Ahmad, Z.; Ghafoor, A. Cowpea aphid-borne mosaic potyvirus: A review. *Int. J. Pest Manage.,* **2002**, *48*(2), 155-168.
[http://dx.doi.org/10.1080/09670870110118722]

[63]   Freitas, J.C.O.; Viana, A.P.; Santos, E.A.; Silva, F.H.L.; Paiva, C.L.; Rodrigues, R.; Souza, M.M.; Eiras, M. Genetic basis of the resistance of a passion fruit segregant population to Cowpea aphid-borne mosaic virus (CABMV). *Trop. Plant Pathol.,* **2015**, *40*(5), 291-297.
[http://dx.doi.org/10.1007/s40858-015-0048-2]

[64]   Salaudeen, M.T.; Aguguom, A. Identification of some cowpea accessions tolerant to cowpea mild mottle virus. *Int. J. Sci. Nat.,* **2014**, *5*(2), 261-267.

[65]   Byamukama, E.; Robertson, A.E.; Nutter, F.W., Jr *Bean pod mottle virus* time of infection influences soybean yield, yield components, and quality. *Plant Dis.,* **2015**, *99*(7), 1026-1032.
[http://dx.doi.org/10.1094/PDIS-11-14-1107-RE] [PMID: 30690975]

[66]   Smith, C.M.; Gedling, C.R.; Wiebe, K.F.; Cassone, B.J. A sweet story: Bean pod mottle virus transmission dynamics by Mexican bean beetles *(Epilachna varivestis). Genome Biol. Evol.,* **2017**, *9*(3), 714-725.
[http://dx.doi.org/10.1093/gbe/evx033] [PMID: 28204501]

[67]   Bradshaw, J.D. Bean pod mottle virus biology and management in Iowa. In: *Doctor of Philosophy*; Iowa State University, Digital Repository, **2007**; p. 7051659. Available from: https://lib.dr.iastate.edu/rtd/15938/ (cited 2021 Sep 27).

[68]   Rehman, F.U.; Kalsoom, M.; Adnan, M.; Naz, N.; Ahmad Nasir, T.; Ali, H. Soybean mosaic disease (SMD): A review. *Egypt J Basic Appl Sci.,* **2021**, *8*(1), 12-16.

[69]   Widyasari, K.; Alazem, M.; Kim, K.H. Soybean resistance to soybean mosaic virus. *Plants,* **2020**, *9*(2), 219.
[http://dx.doi.org/10.3390/plants9020219] [PMID: 32046350]

[70]   Kebede, M.; Ayana, B. Economically important parasitic weeds and their management practices in crops. *J. Environ. Earth Sci.,* **2018**, *8*(12), 104-115.

[71]   Boukar, O.; Togola, A.; Chamarthi, S.; Belko, N.; Ishikawa, H.; Suzuki, K. Cowpea *[Vigna unguiculata (L.) Walp.]* Breeding. In: *Advances in Plant Breeding Strategies: Legumes*; Al-Khayri, JM.; Jain, SM.; Johnson, DV., Eds.; Springer International Publishing: Cham, **2019**; pp. 201-243. Available from: http://link.springer.com/10.1007/978-3-030-23400-3_6 (cited 2021 Sep 20).

[72]   Togola, A.; Boukar, O.; Belko, N.; Chamarthi, S.K.; Fatokun, C.; Tamo, M.; Oigiangbe, N. Host plant resistance to insect pests of cowpea *(Vigna unguiculata L. Walp.)* : achievements and future prospects. *Euphytica,* **2017**, *213*(11), 239.
[http://dx.doi.org/10.1007/s10681-017-2030-1]

[73] Ba, N.M.; Huesing, J.E.; Dabiré-Binso, C.L.; Tamò, M.; Pittendrigh, B.R.; Murdock, L.L. The legume pod borer, Maruca vitrata Fabricius (Lepidoptera: Crambidae), an important insect pest of cowpea: A review emphasizing West Africa. *Int. J. Trop. Insect Sci.,* **2019**, *39*(2), 93-106.
[http://dx.doi.org/10.1007/s42690-019-00024-7]

[74] Amusa, DO; Ogunkanmi, AL; Bolarinwa, K; Ojobo, O Evaluation of four cowpea lines for bruchid *(Callosobruchus maculatus)* to tolerance. *J Nat Sci Res,* **2014**, *3*(13), 46-52.

[75] Koch, R.L.; Sezen, Z.; Porter, P.M.; Ragsdale, D.W.; Wyckhuys, K.A.G.; Heimpel, G.E. On-farm evaluation of a fall-seeded rye cover crop for suppression of soybean aphid *(Hemiptera: Aphididae)* on soybean. *Agric. For. Entomol.,* **2015**, *17*(3), 239-246.
[http://dx.doi.org/10.1111/afe.12099]

[76] Silva, J.P.G.F.; Baldin, E.L.L.; Souza, E.S.; Canassa, V.F.; Lourenção, A.L. Characterization of antibiosis to the redbanded stink bug *Piezodorus guildinii (Hemiptera: Pentatomidae)* in soybean entries. *J. Pest Sci.,* **2013**, *86*(4), 649-657.
[http://dx.doi.org/10.1007/s10340-013-0527-5]

[77] Fourie, H.; Mc Donald, A.H.; Steenkamp, S.; De Waele, D. Nematode pests of leguminous and oilseed crops. In: *Nematology in South Africa: A View from the 21st Century*; Fourie, H.; Spaull, V.W.; Jones, R.K.; Daneel, M.S.; De Waele, D., Eds.; Springer International Publishing: Cham, **2017**; pp. 201-230. [Internet] Available from: http://link.springer.com/10.1007/978-3-319-44210-5_9
[http://dx.doi.org/10.1007/978-3-319-44210-5_9]

[78] Khan, M.R.; Sharma, R.K. Fusarium-nematode wilt disease complexes, etiology and mechanism of development. *Indian Phytopathol.,* **2020**, *73*(4), 615-628.
[http://dx.doi.org/10.1007/s42360-020-00240-z]

[79] Kim, K.S.; Vuong, T.D.; Qiu, D.; Robbins, R.T.; Grover Shannon, J.; Li, Z.; Nguyen, H.T. Advancements in breeding, genetics, and genomics for resistance to three nematode species in soybean. *Theor. Appl. Genet.,* **2016**, *129*(12), 2295-2311.
[http://dx.doi.org/10.1007/s00122-016-2816-x] [PMID: 27796432]

[80] Mitchum, M.G. Soybean resistance to the soybean cyst nematode Heterodera glycines : An Update. *Phytopathology,* **2016**, *106*(12), 1444-1450.
[http://dx.doi.org/10.1094/PHYTO-06-16-0227-RVW] [PMID: 27392178]

[81] Kandel, Y.R.; Wise, K.A.; Bradley, C.A.; Chilvers, M.I.; Byrne, A.M.; Tenuta, A.U.; Faghihi, J.; Wiggs, S.N.; Mueller, D.S. Effect of soybean cyst nematode resistance source and seed treatment on population densities of *Heterodera glycines*, sudden death syndrome, and yield of soybean. *Plant Dis.,* **2017**, *101*(12), 2137-2143.
[http://dx.doi.org/10.1094/PDIS-12-16-1832-RE] [PMID: 30677377]

[82] Coyne, D.L.; Cortada, L.; Dalzell, J.J.; Claudius-Cole, A.O.; Haukeland, S.; Luambano, N.; Talwana, H. Plant-parasitic nematodes and food security in sub-saharan africa. *Annu. Rev. Phytopathol.,* **2018**, *56*(1), 381-403.
[http://dx.doi.org/10.1146/annurev-phyto-080417-045833] [PMID: 29958072]

[83] Jones, J.T.; Haegeman, A.; Danchin, E.G.J.; Gaur, H.S.; Helder, J.; Jones, M.G.K.; Kikuchi, T.; Manzanilla-López, R.; Palomares-Rius, J.E.; Wesemael, W.M.L.; Perry, R.N. Top 10 plant-parasitic nematodes in molecular plant pathology. *Mol. Plant Pathol.,* **2013**, *14*(9), 946-961.
[http://dx.doi.org/10.1111/mpp.12057] [PMID: 23809086]

[84] Kankam F, Sowley E, Dankwa IN. Management of root-knot nematode *(Meloidogyne incognita)* on cowpea *(Vigna unguiculata L. Walp.)* with oil cakes. *Int. J. Biosci.,* **2014**, *5*(12), 413-419.
[http://dx.doi.org/10.12692/ijb/5.12.413-419]

CHAPTER 3

Indexing for Bacterial, Fungal and Viral Pathogens in Legume Plants

Phumzile Mkhize[1,*], Josephine Malatji[1] and **Phetole Mangena[2]**

[1] Department of Microbiology, Biochemistry and Biotechnology, School of Molecular and Life Sciences, Faculty of Science and Agriculture, University of Limpopo, Limpopo Province, Republic of South Africa

[2] Department of Biodiversity, School of Molecular and Life Sciences, Faculty of Science and Agriculture, University of Limpopo, Limpopo Province, Republic of South Africa

Abstract: Microorganisms found in plants exist as epiphytes or endophytes. Most epiphytes remain on plant surfaces and the latter may be intracellular pathogens, opportunistic and adapted microbial colonisers that originate from the surrounding environment. The main purpose of agricultural practices is thus, to develop disease-free varieties by propagating plants under controlled environmental conditions. Such conditions should be optimal for plant production and reduce disease development. The former requires strict certification schemes *via* several routes that include (i) indexing with subsequent removal of infected or contaminated materials from the production chain (ii) meristem and other tissue culture production systems and (iii) the use of thermo or chemotherapy for phytosanitation. Other methods also require balancing and proper adjustments in fertilizer usage and crop rotation. Therefore, this chapter reviews the role of microbial pathogen indexing as a means of controlling bacterial, fungal, and viral diseases that have a significant role to play in agriculture.

Keywords: Agriculture, Bacterial diseases, Fungal diseases, Indexing, Microorganisms, Microbial pathogens, Viral diseases.

INTRODUCTION

Microorganisms found in plants exist as epiphytes or endophytes. Most epiphytes remain on the plants' surface and the latter may be intracellular pathogens, opportunistic and adapted microbial colonisers that originate from the surrounding environment [1]. The adverse effect of the presence of bacterial infections in most legumes cannot go unnoticed as they affect growth, and cause leaf spots, specks and blights, galls, and cankers [2]. Among the bacterial diseases in plants, those that are caused by gram-negative bacteria are the most widespread and destructive. The bacteria of the genus *Pseudomonas, Ralstonia, Agrobacterium, Xanthomonas, Erwinia, Xylella, Pectobacterium,* and *Dickeya* [3] are among the

---

* Corresponding author Phumzile Mkhize: Department of Microbiology, Biochemistry and Biotechnology, School of Molecular and Life Sciences, Faculty of Science and Agriculture, University of Limpopo, Limpopo Province, Republic of South Africa; Tel: +2715-268-3017; E-mail: Phumzile.Mkhize@ul.ac.za

Phetole Mangena & Sifau A. Adejumo (Eds.)

most disruptive genus causing great losses to the agricultural industry. They have a broad crop host range that includes leguminous and non-leguminous crops such as cotton, rice, cereals walnut, soybean, and sugarcane.

These pathogens affect different plant parts including the leaves, stems, and fruits. Unlike viruses, most bacterial plant diseases do not require insects as a vector, but rather rain, wind, soil, seed dispersal or any other means of transport to gain entry into the plants. Microbial pathogens are generally eliminated by chemical microbicides that usually contain copper in combination with ethylene bis-dithiocarbamate like mancozeb, streptomycin and oxytetracycline [4]. Approximately 300,000 species of flowering plants that include cereals, lumber, pulses, barley, corn, rice, sorghum, wheat, and nuts house over 100,000 fungal species. Those include fungal species such as *Magnaporthe oryzae, Botrytis cinerea, Puccinia* spp., *Fusarium graminearum, Fusarium oxysporum, Blumeria graminis, Mycosphaerella graminicola, Colletotrichum* spp., *Ustilago maydis* and *Melampsora lini* [5–8].

These may be pathogenic and saprophytic fungi that cause the development of several diseases including anthracnose, botrytis rots, downy mildew, *Fusarium* rots, powdery mildews, rusts, *Rhizoctonia* rots, *Sclerotinia* rots and *Sclerotium* rots [9]. These microbes cause deterioration in the growth, yield, and quality of crops, and often result in the utter destruction of superior varieties that are much more valuable for agriculture [10]. For example, dramatic losses of revenue estimated at over 11 million US dollars per year as a result of low seedling survival rates caused by *Fusarium circinatum* have been recorded as reported by Storer *et al.* [10]. Generally, all bacterial, viral, and fungal plant pathogens require a wound to gain entry to cause disease development in plants. Plant tissue wounding caused by beetles (*Ips conophthorus, Ernobium,* and *Pissodes nemorensis*) serves as infection sites in mature plants.

In legumes, these microorganisms have evolved mechanisms to actively transverse the plant's outer structural barriers, the cuticle, and the epidermal cell wall structures. For example, fungal pathogens can secrete a cocktail of hydrolytic enzymes, including cutinases, cellulases, pectinases, and proteases to gain entry into tissues [11]. Moreover, these fungal pathogens can easily spread through contaminated planting pots, irrigation water, and supporting mediums [12]. Viral plant infections cause several complex diseases resulting in necrotic cells, tissues or organs and failure for plant organs to develop fully (hypoplasia) causing dwarfing or stunting. Hypoplasia conditions may also cause tissue overgrowths like the formation of crown gall diseases caused by *Agrobacterium* spp. or club root. The most common symptoms of viral infection in plants range from mosaic patterns, chlorotic, yellowing, and leaf rolling to flower deformation [4].

Diseases like the Red clover nepovirus A (RCNVA) remain the most detrimental and cause dramatic effects on plant vigour and yield. Moreover, members of the genera *Ampelovirus, Clostrerovirus,* and *Vitivirus* were also found to be more controversial serving as causal agents for leaf roll and rugose wood. Although these viruses are not much implicated in legume crop diseases; however, they generally cause severe diseases which remain difficult to quantify and estimate because of the complexity related to their mode of transmission and symptoms. Commonly, farmers do not become aware of the real damage until it is too late with the losses culminating in the magnitude of millions of Rands every year [13]. Mixed infection, viral strain, environment, and cultivar response to infection are some of the complex mechanisms of viral plant infections that necessitate specific and accurate indexing methods for their effective control [14].

Indexing for bacterial, fungal and viral pathogens in plants permits the production and use of planting material free from phytopathogenic infections. Even though that remains the case, indexing does not exclusively serve as a method of control and prevention for disease development. The approach remains a vital necessity in the agricultural industry. The main purpose is to develop disease-free varieties by propagating plants under controlled environmental conditions. Such conditions should be optimal for plant production and reduced disease development. The former requires strict certification schemes *via* several routes that include (i) indexing with subsequent removal of infected or contaminated materials from the production chain (ii) meristem and other tissue culture production systems and (iii) the use of thermo or chemotherapy [15]. Furthermore, Pant and Hambly-Odame [16] reported that the latter also requires balancing and proper adjustments in fertilizer usage and crop rotation. Therefore, this chapter reviews the role of microbial pathogen indexing as a means of controlling bacterial, fungal, and viral diseases that have a significant role to play in agriculture, especially leguminous crop production.

INDEXING APPROACHES

The most critical element of developed and optimised indexing systems is that they should not overlook minor infectious microorganisms while focusing on the major ones. Such an approach may be detrimental to the agricultural industry. For example, the impact of minor viruses such as fleck, vein mosaic, and rupestris pitting is amplified by a synergistic negative effect of other major viruses. Furthermore, the mutagenic and revolving nature of microbials may also create a constant spree of emerging new pathogens. The mutagenic rate in viruses is so high such that a single RNA molecule gives rise to a population of mutant sequences (haplotypes or variants) originating from the master sequence (quasispecies). These mutations emanate from the lack of viral polymerase activities required for proofreading of the virus sequences [17 - 19].

The variant can be more disruptive than the mother virus like in case of grape *pinot gris virus*, peach *latent mosaic viroid* [20], *Citrus tristeza virus* [21], and *Citrus exocortis viroid* [22]. If these minor infections remain undetected, they may have as much impact as the major microbial pathogens. Bacteria (such as Proteobacteria as exemplified in Fig. (**1**), and fungal pathogen identification is generally achieved by microscopic examination to identify mycelial morphological characteristics, and the absence/presence of spores, colony appearance as well as pigmentation [23, 24]. Moreover, physical monitoring of disease symptoms development is also important, especially when compared with microscopic examination which may be somewhat tedious [25, 26]. Furthermore, where microscopic examinations face limitations, alternative methods such as the use of selective media may be utilised. For example, malachite green agar 2.5 ppm is one of the developed potent selective media for the isolation and enumeration of *Fusarium* spp.

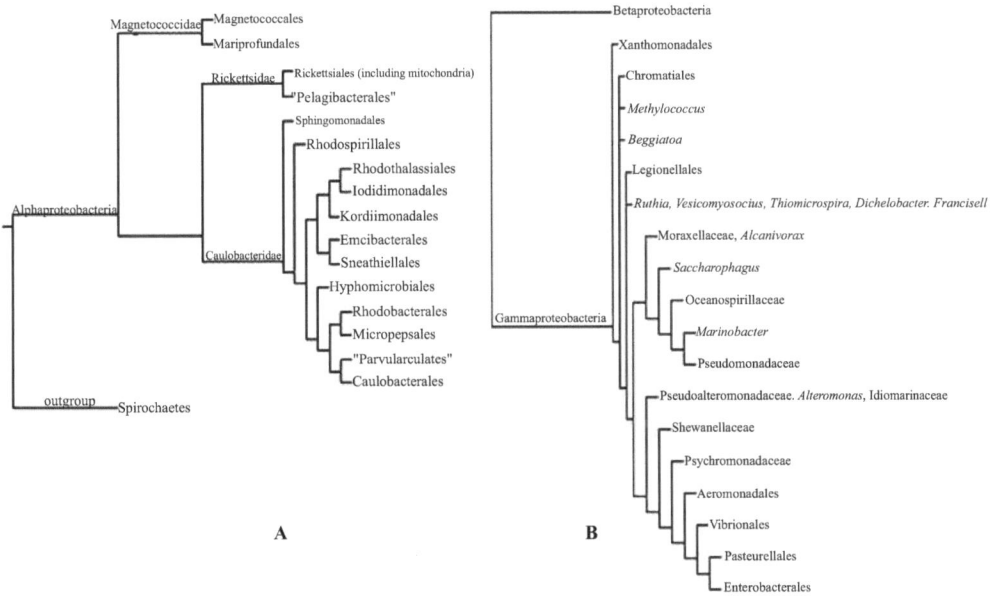

Fig. (1). Phylogeny of a whole analysis of a class of *Alphaproteobacteria* (**A**) and the phylogeny of a whole analysis of a class *Gammaproteobacteria* (**B**) as presented by Williams *et al.* [60].

In addition, potato dextrose agar (PDA), corn meal agar, malt agar and carnation leaf agar serve as some of the most widely used selective media [27]. Bacteriological indexing media is another approach that allows for specific detection of a wide range of bacteria even when bacterial cells are present in low numbers [28]. These can be optimized by preparation of the indexing medium at different concentrations and running multiple cycles [2]. The use of nucleic acid-based detection methods based on polymerase chain reaction (PCR) and protein-based immunochemical techniques is more sensitive, quick, and specific [29 - 31]. These methods, particularly, PCR-based techniques, have been documented for reliable detection of several microbial pathogens and are able to differentiate pathogens up to species level [32 - 34]. Current indexing technologies rely greatly on the in-depth understanding of the different infectious microorganisms. For instance, the research on tobacco mosaic virus (TMV) has greatly permitted the understanding of the different viruses, in the context of transmission, genome size, distribution and overall mechanisms of pathogenicity. The virus has been purified, analysed by electrophoresis, viewed under the microscope for sequencing of the viral coat protein, and elucidated in atomic details [35, 36].

ELEMENTS OF PATHOGENIC INFECTIONS

The mechanisms used by microbial pathogens to cross the protective barrier of plants have been identified. According to the literature, microbial pathogens contain coding regions constituting enzyme complexes facilitating cell-cell movement [37, 38]. Viral and bacterial pathogens contain membrane-bound protein systems responsible for transferring macromolecular virulence factors into host cells. These factors are used to facilitate and carry out host cell infections. For example, *Agrobacterium tumefaciens* and *A. rhizogenes* use mutable and changeable different proteins to transport bacterial DNA into the plant cell nucleus. Moreover, thousands of bacterial microorganisms of different biochemical composition can co-infect plants in a single colony [39]. This clearly indicates that simultaneous detection of these microorganisms may also play a huge role in accelerating their control in causing plant infections. Fungal pathogens contain exo-proteins, glycoproteins and specific proteins produced during the infection process [40].

Furthermore, the diverse biochemical compositions, such as the wide range of secondary metabolites produced by these fungal plant pathogens, which include the terpenes, polyketides, trichothecenes; zearalenone, and enniatins can be used by these organisms to promote infection of plants [41 - 43]. Viruses, bacteria, and fungi use coat proteins, protein enzymes, and mycelium proteins for the infection of plant and animal hosts. These could also serve as potential targets in raising specific and sensitive antibodies for use in diagnostic techniques, such as enzyme-linked immunosorbent assay (ELISA). Similar optimised ELISA-based approaches were reported for rapid and sensitive indexing of red-spotted grouper nervous necrosis virus, beet necrotic yellow vein virus, *potyviruses*, *Bacillus subtilis, Xylella fastidiosa* and *Fusarium* species (*F. circinatum* and *F. oxysporum*) [34, 36, 44].

Members of the Proteobacteria cause diverse disease symptoms, including specks, spots, blights, wilts, galls, and cankers. These bacterial species can cause host-cell deaths in roots, leaves, flowers, fruits, and stems of legumes and other non-leguminous crops. Generally, microbial pathogens invade plant tissues and establish a parasitic relationship between themselves and their hosts [45]. Bacterial mechanism of pathogenicity differs according to species of bacteria whereby some use what is referred to as the type III mechanism of degrading the plant's cell wall material using enzymes such as chitinase, pectinase and cellulases. Cell Wall Degrading Enzymes (CWDEs) are a heterogeneous group of enzymes including glycosyl-hydrolases, oxidoreductases, lyases, and esterases. Once the cell wall is degraded, the bacteria then deposit cellular constituents into the plant cell, thus taking over the host's cellular mechanism. This mechanism is controlled by a cascade of events, including, differential expression of genes that encode for different proteins. Such genomic molecular sequences are:

1. Proteins that comprise the needle complex, transporting substrates from the bacterial cytosol to the extracellular environment,
2. Transport proteins that translocate secreted proteins into host cells,
3. Enzyme proteins that regulate the secretion process, and
4. Proteins that are injected into host cells, called effector proteins [46].

Mold and Fungal Phylogenic Characteristics

Some of the well-studied mold and fungal plant pathogens comprise species of the fungal-like Oomycetes and Leotiomycetes, such as *Phytophthora infestans, Phytophthora ramorum, Phytophthora ultimum* and *Fusarium* spp. Oomycete species are largely and taxonomically found in the order Peronosporales, especially the genus *Phytophthora*. Unlike other species, *P. ultimum* has a different order (Pythiales), family (Pythiaceae) and genus (*Pythium*). Furthermore, all of these species greatly affect wild and cultivated plant species with *P. ultimum* causing serious root rot and a dumping-off in a diversity of non-leguminous and leguminous agronomic crops such as corn, potato, and soybean [47]. The fungal-like eukaryotic microorganism Oomycetes shares similar biological features with fungi. For example, the formation of spores and hyphae is used for reproduction and the release of enzymes or food absorption, respectively.

The large spectrum of environmental conditions and plant hosts that Oomycetes thrive in is also reflected in their phylogenetic diversity. In terms of their life cycles, species like *Phytophthora* and *Pythium* lie dormant as resting spores (oospores or chlamydospores), while others may remain dormant for several years before resuming their reproductive state [48]. Unlike Oomycota, the Zygomycota is a true fungus, as discussed by Richardson [49]. The following genera are also found in these groups in decreasing order; *Aureobasidum, Rhizopus, Mucor, Arthinium, Phoma, Fusarium, Trichoderma,* and *Botrytis*. Additionally, Ascomycota is mainly found in the kingdom Fungi, subkingdom Dikarya with only two phyla. Some important diseases in this phylum are peach leaf curl, powdery mildew, and wilt diseases. Most species in the Ascomycota and Basidiomycota live on plants as saprobes, parasites, or symbionts. However, Ascomycota largely depends on both living and dead plant materials as nutrient sources [50].

Viral and Bacterial Phylogeny and Characteristics

A review of molecular and plant pathology reported that the most common viruses of legume plants particularly include tobacco mosaic virus, tomato spotted wilt virus, soybean yellow leaf curl virus, cucumber mosaic virus, potato virus Y, cauliflower mosaic virus, African cassava mosaic virus, plum pox virus, brome mosaic virus, *Citrus* tristeza virus, barley yellow dwarf virus, leafroll virus, and bushy stunt virus. Some of the most widely studied viruses entail groups of positive sense ssRNA viruses in a family of *Potyviridae* genus; *Potyvirus* with distinct species names officially recognized by the International Committee on Taxonomy of Viruses (ICTV). The most common target host for these viruses is legume crop plants commonly infected by common bean mosaic virus, soybean mosaic virus, pea seed-borne mosaic virus and bean yellow mosaic virus.

Soybean mosaic virus (SMV) infection usually results in severe yield losses and seed quality reduction. It is a seed-borne viral pathogen, and it can be efficiently spread from one plant species to the other by aphids (*Aphis fabae* and *Myzuz pericae*) while feeding on the plant [51]. Over 30% of the seeds from SMV-infected soybean plants carry SMV depending on cultivar and the time of infection before flowering. SMV-infected seeds are the primary inoculum source with weeds and other plants also serving as a reservoir of SMV. The molecular mechanism of *potyviral* long-distance movement is poorly understood due to the complexity involving various cellular structures [52]. Furthermore, SMV often infects soybeans with other viruses such as bean pod mottle virus, alfalfa mosaic virus, and tobacco ringspot virus. In case of pea seed-borne mosaic virus (PSbMV), infections cause disease in pea fields resulting in substantial seed quality and yield losses.

Further research shows that the infection levels are higher early in the life of the subsequent crop [53]. The symptoms induced by these viruses are relatively similar and they depend on host genotype, virus strain, plant age at the time of infection, and the environment. The symptoms commonly observed include the most common green/ yellow mosaic pattern, stunting, leaf curling and seed coat mottling, male sterility, flower deformation, less pubescent, sometimes necrotic local lesions, systemic necrosis, and bud blight. Moreover, Proteobacteria include many bacteria that are part of the normal animal microbiota as well as many plant pathogens (Fig. **1**). Within this group, the many bacteria have a wide variety of metabolic abilities, ecological niches, and cellular characteristics, with most bacterial species within the Alphaproteobacteria carrying out parts of their life cycle in host cells [60].

DIAGNOSTIC METHODS FOR MICROBIAL PLANT PATHOGENS

Disease management relies strongly on a fast and accurate detection of the causal agent. Diagnostic methods for bacterial, fungal, and viral plant pathogens are well documented with some methods conferring specificity to species level, whereas in others, the problem of cross-reactivity and false positive results remains a major challenge [54 - 56]. Methods included in the diagnosis of plant pathogens entail traditional evaluation of symptoms, culturing methods, serological techniques, high-throughput sequencing, nucleic acid-based methods, genomic probes, proteomic approach, biosensors, and multiplexing [57]. Serological methods rely on the specific binding of antibodies to the protein (epitope) of interest expressed by the pathogen [58]. Nucleic acid-based methods rely on the specific binding between nucleic acids dependent on their complementarity [59]. These binding events are visualized by attached markers relying upon factors such as fluorescent dyes, enzyme-based colorimetric or chemiluminescent reactions, and radioactive particles. The selection of which method to use is vastly dependent on the availability of equipment, cost issues, and qualified personnel with specific expertise.

The accuracy of results, consistency, and reproducibility of detection systems are among the paramount factors that need proper consideration for the optimisation and adoption of these detection tools. The general plant microbial pathogen isolation protocol for diagnosis is relatively similar between different diseases. Isolation is carried out from an infected plant material/ seed collected from the field. The material is then surface sterilized using 2% sodium hypochlorite, 95% alcohol or sterile distilled water depending on the diagnostic method. In case of the use of the selective medium, the material will be inoculated on the selective medium; for example, culture media, such as yeast dextrose calcium carbonate and Kelman's triphenyl tetrazolium chloride media, which are some of the most used media by researchers [21, 61, 62].

Colony or mycelium formation is monitored and analysed based on size, shape, and colour and may be subjected to microscopic analysis or biochemical/physiological properties, hypersensitive, and pathogenicity tests for confirmation of the pathogen [63]. For DNA extraction or protein isolation, the plant or seed material is homogenised in liquid nitrogen to powder followed by the addition of buffer, and then centrifugation. From the supernatant, serological methods are conducted, and DNA isolation kits are generally used in case of nucleic acid-based diagnosis [55, 56, 64, 65]. Some researchers use the plant material to subculture the pathogen, and thereafter conduct extraction from the growth medium. During the isolation step, the distinct separation of plant protein and nucleic acid remains a major problem for most researchers [66].

PCR-Based Detection

Polymerase chain reaction (PCR) is the nucleic acid-based detection method that allows for the amplification of a specific region of the genome to make millions of copies. It is used in the early stages of processing DNA for sequencing and for detecting the presence or absence of a gene to help identify pathogens during infection. This reaction is based on the binding of a specific primer/s that only reacts with the target sequence of interest. The basic stages of a PCR include:

1. Denaturing– when the double-stranded template DNA is heated to separate it into two single strands,
2. Annealing– when the temperature is lowered to enable the DNA primers to attach to the template DNA,
3. Extending– when the temperature is raised and the new strand of DNA is made by the Taq polymerase enzyme, and
4. Visualising– the reaction is visualized by electrophoresis or by hybridization with fluorescence probes.

For more specific identification, the amplified product can further be characterised by Sanger sequencing (first-generation sequencing) where it is compared with the known sequences from databases such as Gene Bank, National Centre for Biotechnology Information (NCBI), Nucleotide Sequence Database Collaboration at the European Bioinformatics Institute (EBI) and MycoBank. Real-time PCR works in a similar manner as the general PCR however, in the former, the reaction progress is monitored by the fluorescent reporter that binds to the dsDNA or is released from sequence-specific probes of 15 to 30 nucleotides [67]. Co-operational amplification (Co-PCR) is another method that has been used for the detection of microbial pathogens.

It is more sensitive than the general PCR and over 100 times more sensitive when compared to real-time PCR as it allows the tetra-primer reaction based on the simultaneous amplification from all the primers [68]. The reaction process consists of the simultaneous reverse transcription of two different fragments from the same target, one containing the other. These include the production of four amplicons by the combination of the two pairs of primers whereby one pair is external to the other and the cooperative action of amplicons to produce the largest fragment [69].

PCR-SSCP and Competitive PCR

PCR-single-strand conformation polymorphism (PCR-SSCP) is a simple and powerful technique for identifying sequence changes in amplified DNA. This tool is important for microbial pathogen detection as it can identify possible mutations or new variants or strains of a pathogen in the field. Furthermore, this technique follows the PCR amplification of targeted sequences wherein the amplified product is denatured into single-stranded DNAs (ssDNA) and then subjected to a non-denaturing polyacrylamide gel electrophoresis [70]. Under non-denaturing conditions, ssDNA has a secondary structure that could be determined by the nucleotide sequence. The mobility of the ssDNA depends on the secondary structure of the amplified product. The different positions of the bands of ssDNA on the gel indicate different sequences [71]. Competitive PCR is the variation of the common PCR where multiple primers of different pathogens are used and a signal is observed for each reaction. There is also immune-capture PCR where the pathogen is first detected by a capture antibody and thereafter amplified by PCR. These nucleic acid-based methods are generally expensive and require the use of specialized equipment and expertise, which can be limiting to low-income users. However, they remain the most specific and reliable methods in the detection of most plant pathogens.

Serological Methods of Pathogen Detection

Immunofluorescence, immunoblotting, enzyme-linked immunosorbent assays (ELISA), modified ELISA and dip-stick assays are some of the serological diagnostic methods successfully used in the detection of microbial plant pathogens [40, 66, 72]. ELISA tests have an advantage over the other PCR assays, selection media, and microscopy because they require little sample preparation which makes them suitable for screening many samples, especially from various plant tissue materials. However, with antigens isolated from plant extracts, non-specific reactions due to normal plant constituents need to be considered. This method can be optimized with more sensitive formats such as those of double-sandwich ELISA and monoclonal antibodies with increased specificity [73].

Simultaneous identification of several microorganisms in a single assay is important to increase their feasibility for routine use. Different approaches, based on different biochemical principles are available to simultaneously detect multiple plant diseases. Some of those techniques are multiplex or polyvalent polymerase chain reaction (multiplex-PCR), molecular hybridization and next-generation sequencing (NGS) technologies or what is known as high-throughput sequencing [74]. The use of ELISA-based detection on the antigen-antibody binding principle permits 96 tests to be conducted simultaneously. Multiple antibodies can be used for testing a variety of antigens in microorganisms with the potential detection of multiple infections in one plant or on-site, in the field [57]. However, the need for on-site detection may limit the use of such a technique, accompanied by low sensitivity and the need to often use passive absorption of proteins on hydrophobic surfaces that may cause undesirable changes in the detected proteins which may result in false positive detection outcomes.

Multiplexing Real Time PCR

Multiplexing RT polymerase chain reaction (PCR) has been optimized for the simultaneous detection of more than 10 plant diseases. In this technique, different sets of specific primers of different targets are incorporated in one tube where simultaneous amplification of different nucleic acids occurs in one tube. As many as nine pathogens have been simultaneously detected using this technique [75 - 77]. However, the sensitivity of such a technique can be reduced by the possibility of primer dimerization, and research has indicated that the use of more than five primers can greatly affect the specificity and sensitivity. Improvements in nucleic acid extraction methods have been reported to greatly overcome such sensitivity issues. The multiplexing potential of next-generation sequencing has shown to be a powerful technology that can generate as many as three billion hits per run with read length that varies from 35 to 8000 nucleotides [78 - 80]. In this method, the total number of microorganisms infecting a crop plant can be detected, together with those that are unknown or previously missed by other biological, serological, and molecular tests [81].

Next Generation Sequencing (NGS)

Next-generation sequencing (NGS) allows simultaneous detection and identification of viruses belonging to different families and genera. But can also identify multiple isolates or variants of the same virus co-infecting a single host [82]. NGS involves extracting DNA/RNA, fragmenting it into multiple pieces to add adapters, followed by parallel sequencing of millions of small fragments [83]. Massively parallel signature sequencing, pyrosequencing, polony sequencing, and sequencing by oligonucleotide ligation detection are some of the commonly available advanced sequencing methods in NGS [84]. Bioinformatics analyses are used to connect together these individual reads, compared to the reference genome. This is repeated multiple times, providing accuracy and insights into unexpected RNA/DNA variations [85]. Using this approach Kreuze *et al.* [86] simultaneously identified the sweet potato feathery mottle virus (SPFMV, family *Potyviridae*), and sweet potato chlorotic stunt virus (family *Closteroviridae*) in co-infected plants. Moreover, they unexpectedly detected distinct badnaviruses (family *Caulimoviridae*) and a new mastrevirus (family *Geminiviridae*). Similarly, Verdin *et al.* [87] detected positive sense and negative sense ssRNA viruses, dsRNA viruses, dsDNA viruses, ssDNA viruses, and a viroid infecting ornamental plants. Unique single polymorphism for the isolates of *Calonectria pseudonaviculata* was also established with NGS.

Comparative genomics using NGS identified seven specific regions in *Pseudoperonospora cubensis* that allowed for the development of diagnostic markers [88]. Over 25 variants of the apple stem pitting virus were simultaneously detected in a single sample, an indication of NGS's potential to detect and differentiate between closely related genomic sequences [80]. However, the availability of suitable bioinformatics tools and expertise to extract the required information from the enormous amounts of data generated may limit the use of NGS [78]. In addition, there is a possibility that the sequence detected may be the remnants of sequences incorporated into the host genome [89]. Moreover, the difficulty in using the NGS method lies in finding the direct association between the disease and a particular virus within the identified viruses infecting one plant. In such cases, the diagnostic method must be supplemented by perhaps fulfilling Koch's postulate method or by finding a tight association between the disease and the presence of a certain microorganism during field surveys [90].

BIOLOGICAL AND CHEMICAL PLANT DISEASE CONTROL

Management and control of microbial pathogens remain a challenge for both small farmers and commercial production fields due to the limited efficacy of the current disease management strategies. Most pathogens have acquired resistance to most antimicrobial chemical applications that were generally effective such as streptomycin, fungicides, and some fixed copper bactericides [91]. Host resistance has not proven durable and may be limited to some strains/variants and not others especially emerging ones [79]. In addition, the variation in host resistance may differ within commercial cultivars and even in natural populations [92]. The continuously acquired resistance to antimicrobial chemicals may be a result of their excessive use. In addition, microbial pathogens acquire resistance to most of these chemicals by changing the target active sites in their enzymatic structure and by active efflux of the "drugs" from the cell walls [93].

Some microorganisms have evolved specific mechanisms to degrade routine antibiotics pertaining to their ability to synthesize hydrolytic enzymes that are encoded by the plasmid genes found freely in their cells [94, 95]. Others have active efflux that involves pumping the "drug" out of their cells giving them the advantage of surviving under such drug application. Lastly, chemical modification of the target site making the control drug less compatible with such a site which results in less efficacy of the "drug" are some of the biochemical changes reported to enhance and promote microbial drug resistance [96]. This necessitates the development of control measures effective enough to destroy the microorganisms chemically and physically by escaping all the defence measures currently in place. Research in finding alternative and most suitable means of controlling microbial pathogens in plants needs to be effective and not harmful to other associated living creatures especially humans and animals and must be environmentally friendly [93]. Biological control measures have received a lot of attention. Bioactive products (*e.g.* Acibenzolar-S-methyl) referred to as plant activators are one of the strategies that have been used to induce synthetic acquired resistance in tobacco and cucumber [97]. This resistance is associated with a cascade of defence reactions including the accumulation of pathogenesis-related proteins and ultra-structural changes [98, 99]. This method has also proven effective against some fungal, oomycete and bacterial diseases including bacterial spots caused by *Xanthomonas axonopodis pv. vesicatoria* [98, 99]. In addition, colloidal silver nanoparticles and their applications as a biological control agent against bacterial pathogens have proved to be effective in controlling *Pectobacterium, Carotovorum, X. oryzae, X. vesicatoria* and *Ralstonia solanacearum* [93].

The antimicrobial activity of these particles is achievable through the interaction of the positive charge on the surface of the nanoparticle with the negative charge on the microbial plasma membrane and nucleic acid. This destabilises the cell membrane and causes the release of reactive oxygen species and the breakdown of DNA structures [100]. These methods are effective and have no negative impacts on the environment; however, there is still a need to improve their efficacy to optimise their adoption. The use of phages for disease control is a fast-expanding area of plant protection with great potential to replace chemical control measures. These viruses only infect and multiply within their specific host, disrupt metabolism and cause bacteria to lyse [101]. Phages have been successfully used as biocontrol agents for phytobacteria, bacterial wilt of tomato and bacterial leaf blight disease caused by *Xanthomonas axonopodis pv. Allii*. Some fundamental considerations needed during their formulation include their stability, the production time and cost, the impact they might have on untargeted bacteria as well as ideal timing for their optimal application as biocontrol agents.

Several strains of fungus *Trichoderma* have been isolated and found effective in controlling bacterial diseases both in the greenhouses and field conditions. Their mode of action is through the recognition by lectins between the *Trichoderma* and its host fungi. This is followed by the action of degradation enzymes such as chitinase and β-1,3 glucanases that target the cell wall of the host leading to the release of the cell constituents and subsequently causing death [102] Similarly, *Rhizobium* and *Pseudomonas* reduced the severity of bacterial pathogens such as bacterial leaf blight of rice. Evaluating and optimising the combined application of these biocontrol agents by studying additive, synergistic and antagonistic effects of their combinations could optimize their usage and effectiveness.

FUTURE PROSPECTS

Development of new diagnostic methods and improvement of the currently available ones require in-depth knowledge of the targeted microorganisms. This is important, particularly, for the selection of specialized molecules that can be targeted through those newly developed detection methods. Perhaps, common biochemical molecules found in microorganisms of the same family or genus could be important whereby diagnosis done at such level might yield very significant results. Similarly, pathogen detection that is done at a specific species level requires the identification of unique molecules found in such species. For example, plant viruses have a general nomenclature common across the different taxa, however, this may differ for viruses that infect humans and animals. Some are single-stranded positive sense RNA that form rod-shape virions which have an open reading frame that is translated into different proteins [103]. Cell wall molecules such as α-glucan, β-glucan, chitin, galactomannan, and melanin can be potentially used as targets for the development of newly sophisticated, reproducible and specific diagnostic techniques with high specificity to species level.

The different types and functions of proteins may also provide quicker diagnosis, identifying unique potential targets during proteomic analysis and enzymatic assay analysis. For example, the presence of cysteine proteases is easily characterised by the use of two-dimensional gel electrophoresis with an expected size of approximately 25 to 30 kDa at pH 3 to pH 4. In this case, cystatin or E-64 which acts as an inhibitor of the protease enzyme activity, can be used for their characterisation [104]. Moreover, such molecules contain areas of conserved motifs that are important for the detection of potential mutant viruses. For instance, RNA capping methyltransferase, nucleotide triphosphatase (NTPase) or helicase domains and the RNA polymerase domain identified in the *Brome mosaic virus* (BMV) genome were found to be highly conserved, serving as potential targets for specific detection of this virus [105].

Some viruses may have additional sites distinguishing them from the common nomenclature. Such uniqueness may occur through the shift in the open reading frames that give rise to the varying amounts of conserved regions between viruses of different genera. The minority of plant viruses that possess dsDNA and dsRNA genomes, and those containing negative sense RNA such as bacilliform plant viruses are possible targets for specified diagnosis with nucleic acid-based detection methods [106]. Some viruses have rigid rods whereas others are flexuous filaments or icosahedral or icosahedra and bacilliform, which serve as important parameters for microscopic distinguishment. In some of the studies, the existence of a 32 KDa P1 protein in tobacco plants infected with potato Y ordinary strain (PVY-O) was detected by antiserum against a recombinant PVY-OP1 protein [107]. Moreover, HC-Pro and CP have been detected using western blot in the same plants.

The P1 protein was only detectable during the symptom development indicating its relatively short life. This fast turnover of P1 is considered a post-translational modification of gene expression, which needs to be considered in viral diagnostic methods [108]. Furthermore, more studies have successfully used immunological methods for the detection of plant viruses where different proteins were targeted [109]. Such biochemical properties permit continuous improvement of different diagnostic methods and can also be used to guide future studies that are aimed at developing new on-site pathogen detection systems.

CONCLUDING REMARKS

The paramount obstacles in the accurate diagnosis of microbial pathogens comprise the following: (i) the diverse host range that exists both in the species complex and within a single species leading to misidentifications/false positive diagnostic results, (ii) lack of reliably accurate evidence-based molecular diagnostic tools and better biomarkers, (iii) asymptomatic hosts indicating large quantities of pathogen DNA, and (iv) many countries with no stringent import regulations on mycotoxin contamination, and possibly permitting the introduction of new fungal variants into new areas [7, 110]. The research available on plant pathogens permits the development of diagnostic methods with improved sensitivity and specificity. For example, a complete genome of *Aspergillus*, *Condida* and *Saccharomyces* pathogens is available, and importantly used for accelerating gene-based fungal detection and possible development of markers for the diagnosis of these pathogens [111].

Moreover, some fungal pathogens do not share a common set of genes with others as in the case of *Aspergilus fumigatus* and *Fusarium oxysporum* [111, 112]. The common molecular features of fungal conidia and hyphae are one of the hindering characters affecting the development of species-specific diagnostic methods. These may include glycoprotein scaffolds, carbohydrate epitopes, and cell wall materials of the fungus, especially chitin as described by Wycoff *et al.* [113], De-Bernardis *et al.* [114] and Hitchcock *et al.* [115]. However, the fungal cell wall is a three-dimensional structure consuming about one-quarter of fungal cell biomass. As previously indicated, the possible targets for fungal identification can be the use of chitin, a composition of the cell wall, which is different between genera and among species of the same genus. Therefore, with optimized protein isolation techniques obtaining species-specific antibodies, detection is possible. It must be noted that antibodies raised may show inconsistencies in their reactivity as fungal pathogens express different structural proteins during their life cycle [116].

In addition, these protein changes may affect the detection levels by both polyclonal and monoclonal antibodies as slight changes in epitope structure led to drastic reductions in the antibody binding capacity. Such common biochemical features indicate the need for continuous research where species specificity is required. The basic pathogenic features can be targeted in the development and improvement of the currently available detection methods and those aiming for detection at the genus level. For example, the outer conidial surface contains protrusions, termed rodlets whose expression is controlled by the *Atg5* gene [117]. Moreover, the surface-exposed sialic acid residues could be targeted in affinity purification methods (Wasylnka and Moore, 2000; Warwas *et al.*, 2007). The cell wall consists predominately of a polysaccharide matrix comprised of α-(1,3) glucan, β-(1,3) glucan, some of which contains β-(1,6) branches, linear β-(1,3), β-(1,4) glucan, chitin, and galactomannan [118].

The conidial cell wall also contains at least nine glycosylphosphatidylinositol (GPI)--linked proteins connected to the polysaccharide skeleton [119]. These could be potential targets in diagnostic methods such as protein mass spectrometry and two-dimensional electrophoresis (2D- electrophoresis) [isoelectric focusing (IEF) sodium-dodecyl polyacrylamide gel electrophoresis (SDS-PAGE)] which make it possible to identify unique and specific proteins with the potential to produce highly specific antibodies [120, 121]. Methods of obtaining pure and unique proteins in microbial pathogens ensure that antibodies of the best specificity are obtained, especially ensuring the efficacy of the different serological tests. Plant bacterial diagnosis is achievable through different methods that rely on their biochemical properties, and nutritional and metabolic capabilities. Some bacterial pathogens store granules within their cytoplasm that distinguish them from others. For example, poly-β-hydroxybutyrate (PHB) is a carbon- and energy-storage compound found in some nonfluorescent bacteria of the genus *Pseudomonas*. Different species within this genus can be classified by the presence or absence of PHB and fluorescent pigments [1]. Different metabolic tests such as the carbon utilization, pH, chemical sensitivity and lipid profile are the effective methods to diagnose different plant bacteria as these vary from one genus to another even to species level. For example, fatty acids of lipids can vary in chain length, presence or absence of double bonds, and number of double bonds, hydroxyl groups, branches, and rings.

In such instances, phospholipid-derived fatty acids (PLFA), fatty acid methyl ester (FAME) analysis and gas chromatography (GC) analysis can be used for their diagnosis. Methods for easy isolation of these molecules are important in fast-tracking their diagnosis thus ensuring the planting of disease-free crops. More effective, specific, and sensitive techniques are capable of identifying bacteria by determining the specimen's mass spectrum and then comparing it to a database that contains known mass spectra for thousands of microorganisms [122]. The mass spectrum is generated by subjecting the isolated bacterial mixture to irradiation with a high-intensity pulsed ultraviolet laser, resulting in the ejection of gaseous ions generated from the various chemical constituents of the microorganism [123]. These gaseous ions are collected and accelerated through the mass spectrometer, with ions traveling at a velocity determined by their mass-to-charge ratio (m/z), thus, reaching the detector at different times. A plot of detector signal versus m/z yields a mass spectrum for the organism that is uniquely related to its biochemical composition.

Comparison of the mass spectrum to a library of reference spectra obtained from identical analyses of known microorganisms permits the identification of the targeted bacteria. Such detection can be very specific; however, it requires the use of sophisticated equipment that may not be easily available for low-income countries. Similar to viral and fungal disease detection, serological and nucleic acid-based methods are also effective in detecting these pathogens. Fungal diagnosis using serological methods is effective where specific proteins such as the glycoproteins are targeted in raising specific antibodies. The method is effective, but it remains guided by the expertise, cost, and availability of equipment. Research into biochemical molecules that are easily assessable and expressed across different pathogens is important in accelerating their indexing.

LIST OF ABBREVIATIONS

DNA	Deoxyribo Nucleic Acid
EBI	European Bioinformatics Institute
ELISA	Enzyme-linked immunosorbent assay
ICTV	International Committee on Taxonomy of Viruses
NCBI	National Centre for Biotechnology Information
NGS	Next-generation sequencing

kDa	kilo Daltons
PCR	Polymerase Chain Reaction
PCR-SSCP	PCR-single-strand conformation polymorphism
PDA	Potato dextrose agar
PSbMV	Pea seed-borne mosaic virus
RNA	Ribonucleic acid
RT-PCR	Real-time PCR
dsRNA	double-stranded RNA
ssRNA	single-stranded RNA
STV	Soybean mosaic virus
TMV	Tobacco mosaic virus

CONSENT FOR PUBLICATION

Not applicable.

CONFLICT OF INTEREST

The authors declares no conflict of interest, financial or otherwise.

ACKNOWLEDGEMENT

Declared none.

REFERENCES

[1]   Leifert, C.; Waites, W.M. Bacterial growth in plant tissue culture media. *J. Appl. Bacteriol.,* **1992**, *72*(6), 460-466.
       [http://dx.doi.org/10.1111/j.1365-2672.1992.tb01859.x]

[2]   Thomas, P. *In vitro* decline in plant cultures: detection of a legion of covert bacteria as the cause for degeneration of long-term micro propagated triploid watermelon cultures. *Plant Cell Tissue Organ Cult.,* **2004**, *77*(2), 173-179.
       [http://dx.doi.org/10.1023/B:TICU.0000016824.09108.c8]

[3]   Mansfield, J.; Genin, S.; Magori, S.; Citovsky, V.; Sriariyanum, M.; Ronald, P.; Dow, M.; Verdier, V.; Beer, S.; MacHado, M.A.; Toth, I.; Salmond, G.; Foster, G.D. Top 10 plant pathogenic bacteria in molecular plant pathology. *Mol. Plant Pathol.,* **2012**, *13*(6), 614-629.
       [http://dx.doi.org/10.1111/j.1364-3703.2012.00804.x] [PMID: 22672649]

[4]   Buttimer, C.; McAuliffe, O.; Ross, R.P.; Hill, C.; O'Mahony, J.; Coffey, A. Bacteriophages and bacterial plant diseases. *Front. Microbiol.,* **2017**, *8*, 34.
       [http://dx.doi.org/10.3389/fmicb.2017.00034] [PMID: 28163700]

[5]   Rong, I.H.; Baxter, A.P. The South African national collection of fungi: Celebrating a centenary 1905-2005. *Stud. Mycol.,* **2006**, *55*, 1-12.
       [http://dx.doi.org/10.3114/sim.55.1.1] [PMID: 18490968]

[6]   Dean, R.; Van Kan, J.A.L.; Pretorius, Z.A.; Hammond-Kosack, K.; Di Pietro, A.; Spanu, P.D.; Rudd,

J.J.; Dickman, M.; Kahmann, R.; Ellis, J.; Foster, G.D. The Top 10 fungal pathogens in molecular plant pathology. *Mol. Plant Pathol.,* **2012**, *13*(4), 414-430.
[http://dx.doi.org/10.1111/j.1364-3703.2011.00783.x] [PMID: 22471698]

[7]   Liew, W.P.P.; Mohd-Redzwan, S. Mycotoxin: Its impact on gut health and microbiota. *Front. Cell. Infect. Microbiol.,* **2018**, *8*, 60.
[http://dx.doi.org/10.3389/fcimb.2018.00060] [PMID: 29535978]

[8]   Safarieskandari, S.; Chatterton, S.; Hall, L.M. Pathogenicity and host range of *Fusarium* species associated with pea root rot in Alberta, Canada. *Can. J. Plant Pathol.,* **2021**, *43*(1), 162-171.
[http://dx.doi.org/10.1080/07060661.2020.1730442]

[9]   Viljoen, A.; Wingfield, M.J.; Crous, P.W. Fungal pathogens in pinus and eucalyptus seedling nurseries in south africa: A review. *South African Forestry Journal,* **1992**, *161*(1), 45-51.
[http://dx.doi.org/10.1080/00382167.1992.9630424]

[10]   Storer, A.J.; Wood, D.L.; Gordon, T.R. The epidemiology of pitch canker of Monterey pine in California. *For. Sci.,* **2002**, *48*, 646-700.

[11]   Knogge, W. Fungal infection of plants. *Plant Cell,* **1996**, *8*(10), 1711-1722.
[http://dx.doi.org/10.2307/3870224] [PMID: 12239359]

[12]   Baldwin, I.T. Jasmonate-induced responses are costly but benefit plants under attack in native populations. *Proc. Natl. Acad. Sci.,* **1998**, *95*(14), 8113-8118.
[http://dx.doi.org/10.1073/pnas.95.14.8113] [PMID: 9653149]

[13]   Anderson, P.K.; Cunningham, A.A.; Patel, N.G.; Morales, F.J.; Epstein, P.R.; Daszak, P. Emerging infectious diseases of plants: Pathogen pollution, climate change and agrotechnology drivers. *Trends Ecol. Evol.,* **2004**, *19*(10), 535-544.
[http://dx.doi.org/10.1016/j.tree.2004.07.021] [PMID: 16701319]

[14]   Jones, R.A.C. Plant virus emergence and evolution: Origins, new encounter scenarios, factors driving emergence, effects of changing world conditions, and prospects for control. *Virus Res.,* **2009**, *141*(2), 113-130.
[http://dx.doi.org/10.1016/j.virusres.2008.07.028] [PMID: 19159652]

[15]   Elena, S.F.; Fraile, A.; García-Arenal, F. Evolution and emergence of plant viruses. *Adv. Virus Res.,* **2014**, *88*, 161-191.
[http://dx.doi.org/10.1016/B978-0-12-800098-4.00003-9] [PMID: 24373312]

[16]   Pant, L.P.; Hambly-Odame, H. Innovations systems in renewable natural resources management and sustainable agriculture: A literature review. *Afr J Sci Technol Dev,* **2009**, *1*, 103-135.

[17]   Domingo, E.; Escarmís, C.; Sevilla, N.; Moya, A.; Elena, S.F.; Quer, J.; Novella, I.S.; Holland, J.J. Basic concepts in RNA virus evolution. *FASEB J.,* **1996**, *10*(8), 859-864.
[http://dx.doi.org/10.1096/fasebj.10.8.8666162] [PMID: 8666162]

[18]   Drake, J.W.; Holland, J.J. Mutation rates among RNA viruses. *Proc. Natl. Acad. Sci.,* **1999**, *96*(24), 13910-13913.
[http://dx.doi.org/10.1073/pnas.96.24.13910] [PMID: 10570172]

[19]   Gago, S.; Elena, S.F.; Flores, R.; Sanjuán, R. Extremely high mutation rate of a hammerhead viroid. *Science,* **2009**, *323*(5919), 1308.
[http://dx.doi.org/10.1126/science.1169202] [PMID: 19265013]

[20]   Ambrós, S.; Hernández, C.; Flores, R. Rapid generation of genetic heterogeneity in progenies from individual cDNA clones of peach latent mosaic viroid in its natural host. *J. Gen. Virol.,* **1999**, *80*(8), 2239-2252.
[http://dx.doi.org/10.1099/0022-1317-80-8-2239] [PMID: 10466824]

[21]   Kong, P.; Rubio, L.; Polek, M.; Falk, B.W. Population structure and genetic diversity within California Citrus tristeza virus (CTV) isolates. *Virus Genes,* **2000**, *21*(3), 139-145.
[http://dx.doi.org/10.1023/A:1008198311398] [PMID: 11129629]

[22] Gandía, M.; Rubio, L.; Palacio, A.; Duran-Vila, N. Genetic variation and population structure of an isolate of *Citrus exocortis* viroid (CEVd) and of the progenies of two infectious sequence variants. *Arch. Virol.,* **2005**, *150*(10), 1945-1957.
[http://dx.doi.org/10.1007/s00705-005-0570-5] [PMID: 15959832]

[23] Morretti, A.N. Taxonomy of *Fusarium* genus: A continuous fight between lumbers and splitters. *Prog. Nat. Sci.,* **2009**, *117*, 7-13.

[24] Hafizi, R.; Salleh, B.; Latiffah, Z. Morphological and molecular characterization of *Fusarium solani* and *F. oxysporum* associated with crown disease of oil palm. *Braz. J. Microbiol.,* **2013**, *44*(3), 959-968.
[http://dx.doi.org/10.1590/S1517-83822013000300047] [PMID: 24516465]

[25] Houpikian, P.; Raoult, D. Traditional and molecular techniques for the study of emerging bacterial diseases: one laboratory's perspective. *Emerg. Infect. Dis.,* **2002**, *8*(2), 122-131.
[http://dx.doi.org/10.3201/eid0802.010141] [PMID: 11897062]

[26] Luchi, N.; Ioos, R.; Santini, A. Fast and reliable molecular methods to detect fungal pathogens in woody plants. *Appl. Microbiol. Biotechnol.,* **2020**, *104*(6), 2453-2468.
[http://dx.doi.org/10.1007/s00253-020-10395-4] [PMID: 32006049]

[27] Thompson, R.S.; Aveling, T.A.S.; Blanco Prieto, R. A new semi-selective medium for *Fusarium graminearum, F. proliferatum, F. subglutinans* and *F. verticillioides* in maize seed. *S. Afr. J. Bot.,* **2013**, *84*, 94-101.
[http://dx.doi.org/10.1016/j.sajb.2012.10.003]

[28] Casselis, A.C. Contamination detection and elimination in plant cell culture. In: *Encyclopedia of Cell Technology*; Spier, R.E., Ed.; John Wiley: New York, **2000**; 2, pp. 577-5.

[29] Holzhauser, T.; Wangorsch, A.; Vieths, S. Polymerase chain reaction (PCR) for detection of potentially allergenic hazelnut residues in complex food matrixes. *Eur. Food Res. Technol.,* **2000**, *211*(5), 360-365.
[http://dx.doi.org/10.1007/s002170000152]

[30] Janssen, K.; Knez, K.; Spasic, D.; Lammertyn, J. Nucleic acids for ultra-sensitive protein detection. *Sensors,* **2013**, *13*(1), 1353-1384.
[http://dx.doi.org/10.3390/s130101353] [PMID: 23337338]

[31] Sue, M.J.; Yeap, S.K.; Omar, A.R.; Tan, S.W. Application of PCR-ELISA in molecular diagnosis. *BioMed Res Int.,* **2014**, *653014*, 1-6.

[32] Weiland, J.J.; Sundsbak, J.L. Differentiation and detection of sugar beet fungal pathogens using PCR amplification of actin coding sequences and the ITS region of the rRNA gene. *Plant Dis.,* **2000**, *84*(4), 475-482.
[http://dx.doi.org/10.1094/PDIS.2000.84.4.475] [PMID: 30841173]

[33] Maresi, G.; Luchi, N.; Pinzani, P.; Pazzagli, M.; Capretti, P. Detection of Diplodia pinea in asymptomatic pine shoots and its relation to the Normalized Insolation index. *For. Pathol.,* **2007**, *37*(4), 272-280.
[http://dx.doi.org/10.1111/j.1439-0329.2007.00506.x]

[34] Mkhize, P.; Mangena, P. Enzyme immunosorbent assay (ELISA) based detection of *Fusariun circinatum* for alleviation of pine seedling wilt. *J. Biotech Res.,* **2021**, *12*, 23-32.

[35] Namba, K.; Pattanayek, R.; Stubbs, G. Visualization of protein-nucleic acid interactions in a virus. *J. Mol. Biol.,* **1989**, *208*(2), 307-325.
[http://dx.doi.org/10.1016/0022-2836(89)90391-4] [PMID: 2769760]

[36] Shukla, D.D.; Ward, C.W. Identification and classification of potyviruses on the basis of coat protein sequence data and serology. *Arch. Virol.,* **1989**, *106*(3-4), 171-200.
[http://dx.doi.org/10.1007/BF01313952] [PMID: 2673154]

[37]   Heinlein, M. Plant virus replication and movement. *Virology,* **2015**, *479-480*, 657-671.
        [http://dx.doi.org/10.1016/j.virol.2015.01.025] [PMID: 25746797]

[38]   Citovsky, V.; Knorr, D.; Schuster, G.; Zambryski, P. The P30 movement protein of tobacco mosaic
        virus is a single-strand nucleic acid binding protein. *Cell,* **1990**, *60*(4), 637-647.
        [http://dx.doi.org/10.1016/0092-8674(90)90667-4] [PMID: 2302736]

[39]   Pennazio, S. Mineral nutrition of plants: A short history of plant physiology. *Riv. Biol.,* **2005**, *98*(2),
        215-236.
        [PMID: 16180194]

[40]   Arie, T.; Hayashi, Y.; Yoneyama, K.; Nagatani, A.; Furuya, M.; Yamaguchi, I. Detection of *Fusarium*
        spp. in plants with monoclonal antibody. *Ann. Phytopathol. Soc. Jpn.,* **1995**, *61*(4), 311-317.
        [http://dx.doi.org/10.3186/jjphytopath.61.311]

[41]   Díaz-Sánchez, V.; Avalos, J.; Limón, M.C. Identification and regulation of fusA, the polyketide
        synthase gene responsible for fusarin production in *Fusarium fujikuroi. Appl. Environ. Microbiol.,*
        2012, *78*(20), 7258-7266.
        [http://dx.doi.org/10.1128/AEM.01552-12] [PMID: 22865073]

[42]   Tian, Y; Tan, Y; Yan, Z; Liao, Y; Chen, J; Boevre, DM; Saeger, SD; Wu, A Antagonistic and
        detoxification potentials of Trichoderma isolates for control of zearalenone (ZEN) producing
        Fusarium graminearum. Front. Microbiol, **2018**, *8*(27/10), 1-11.

[43]   Liuzzi, V.; Mirabelli, V.; Cimmarusti, M.; Haidukowski, M.; Leslie, J.; Logrieco, A.; Caliandro, R.;
        Fanelli, F.; Mulè, G. Enniatin and beauvericin biosynthesis in *Fusarium* species: Production profiles
        and structural determinant prediction. *Toxins,* **2017**, *9*(2), 45-62.
        [http://dx.doi.org/10.3390/toxins9020045] [PMID: 28125067]

[44]   Ferguson, C.M.J.; Booth, N.A.; Allan, E.J. An ELISA for the detection of *Bacillus subtilis* L-form
        bacteria confirms their symbiosis in strawberry. *Lett. Appl. Microbiol.,* **2000**, *31*(5), 390-394.
        [http://dx.doi.org/10.1046/j.1472-765x.2000.00834.x] [PMID: 11069643]

[45]   Carro, L.; Nouioui, I. Taxonomy and systematics of plant probiotic bacteria in the genomic era. *AIMS
        Microbiol.,* **2017**, *3*(3), 383-412.
        [http://dx.doi.org/10.3934/microbiol.2017.3.383] [PMID: 31294168]

[46]   Vidaver, A.K.; Lambrecht, P.A. Bacteria as plant pathogens. In: *Physic Rev J*; , **2004**.
        [http://dx.doi.org/10.1094/PHI-I-2004-0809-01]

[47]   Hargreaves, J.; van West, P. Oomycete-root interactions. In: *Methods in Rhizosphere Biology
        Research. Rhizosphere Biology*; Reinhardt, D.; Sharma, A., Eds.; Springer: Singapore, **2019**; pp. 83-
        103.
        [http://dx.doi.org/10.1007/978-981-13-5767-1_6]

[48]   Rossmann, S.; Lysøe, E.; Skogen, M.; Talgø, V.; Brurberg, M.B. DNA metabarcoding reveals broad
        presence of plant pathogenic oomycetes in soil from internationally traded plants. *Front. Microbiol.,*
        2021, *12*, 637068.
        [http://dx.doi.org/10.3389/fmicb.2021.637068] [PMID: 33841362]

[49]   Richardson, M. The ecology of the Zygomycetes and its impact on environmental exposure. *Clin.
        Microbiol. Infect.,* **2009**, *15* 5, 2-9.
        [http://dx.doi.org/10.1111/j.1469-0691.2009.02972.x] [PMID: 19754749]

[50]   Berbee, M.L. The phylogeny of plant and animal pathogens in the Ascomycota. *Physiol. Mol. Plant
        Pathol.,* **2001**, *59*(4), 165-187.
        [http://dx.doi.org/10.1006/pmpp.2001.0355]

[51]   Chatzivassiliou, E.K. An annotated list of legume-infecting viruses in the light of metagenomics.
        Plants, **2021**, *10*(7), 1413.
        [http://dx.doi.org/10.3390/plants10071413] [PMID: 34371616]

[52]   Cui, X.; Chen, X.; Wang, A. Detection, Understanding and Control of Soybean Mosaic Virus. In: *Soybean - molecular aspects of breeding*; Sudaric, A., Ed.; IntechOpen: London, United Kingdom, **2011**; pp. 335-354.
[http://dx.doi.org/10.5772/14298]

[53]   Congdon, B.S.; Coutts, B.A.; Renton, M.; Banovic, M.; Jones, R.A.C. Pea seed-borne mosaic virus in field pea: Widespread infection, genetic diversity, and resistance gene effectiveness. *Plant Dis.,* **2016**, *100*(12), 2475-2482.
[http://dx.doi.org/10.1094/PDIS-05-16-0670-RE] [PMID: 30686170]

[54]   Sebire, K.; McGavin, K.; Land, S.; Middleton, T.; Birch, C. Stability of human immunodeficiency virus RNA in blood specimens as measured by a commercial PCR-based assay. *J. Clin. Microbiol.,* **1998**, *36*(2), 493-498.
[http://dx.doi.org/10.1128/JCM.36.2.493-498.1998] [PMID: 9466765]

[55]   Wei, C.; Liu, J.; Maina, A.N.; Mwaura, F.B.; Yu, J.; Yan, C.; Zhang, R.; Wei, H. Developing a bacteriophage cocktail for biocontrol of potato bacterial wilt. *Virol. Sin.,* **2017**, *32*(6), 476-484.
[http://dx.doi.org/10.1007/s12250-017-3987-6] [PMID: 29168148]

[56]   Calderon, C.; Ward, E.; Freeman, J.; Foster, S.J.; McCartney, H.A. Detection of airborne inoculum of *Leptosphaeria maculans* and *Pyrenopeziza brassicae* in oilseed rape crops by polymerase chain reaction (PCR) assays. *Plant Pathol.,* **2002**, *51*(3), 303-310.
[http://dx.doi.org/10.1046/j.1365-3059.2002.00721.x]

[57]   Skottrup, P.D.; Nicolaisen, M.; Justesen, A.F. Towards on-site pathogen detection using antibody-based sensors. *Biosens. Bioelectron.,* **2008**, *24*(3), 339-348.
[http://dx.doi.org/10.1016/j.bios.2008.06.045] [PMID: 18675543]

[58]   Clark, M.F.; Adams, A.N. Characteristics of the microplate method of enzyme-linked immunosorbent assay for the detection of plant viruses. *J. Gen. Virol.,* **1977**, *34*(3), 475-483.
[http://dx.doi.org/10.1099/0022-1317-34-3-475] [PMID: 323416]

[59]   Rouhiainen, L.; Laaksonen, M.; Karjalainen, R.; Söderlund, H. Rapid detection of a plant virus by solution hybridization using oligonucleotide probes. *J. Virol. Methods,* **1991**, *34*(1), 81-90.
[http://dx.doi.org/10.1016/0166-0934(91)90123-H] [PMID: 1955492]

[60]   Williams, K.P.; Gillespie, J.J.; Sobral, B.W.S.; Nordberg, E.K.; Snyder, E.E.; Shallom, J.M.; Dickerman, A.W. Phylogeny of Gammaproteobacteria. *J. Bacteriol.,* **2010**, *192*(9), 2305-2314.
[http://dx.doi.org/10.1128/JB.01480-09] [PMID: 20207755]

[61]   Kelman, A. The relationship of pathogenicity in *Pseudomonas solanacearum* to colony appearance on a tetrazolium medium. *Phytopathology,* **1954**, *51*, 158-161.

[62]   Schaad, N.W. Laboratory Guide for the Identification of Plant Pathogenic Bacteria. *American Phytopathological Society,* **1988**, *5*, 81-82.

[63]   Lievens, B.; Thomma, B.P.H.J. Recent developments in pathogen detection arrays: implications for fungal plant pathogens and use in practice. *Phytopathology,* **2005**, *95*(12), 1374-1380.
[http://dx.doi.org/10.1094/PHYTO-95-1374] [PMID: 18943547]

[64]   Budge, G.E.; Shaw, M.W.; Colyer, A.; Pietravalle, S.; Boonham, N. Molecular tools to investigate *Rhizoctonia solani* distribution in soil. *Plant Pathol.,* **2009**, *58*(6), 1071-1080.
[http://dx.doi.org/10.1111/j.1365-3059.2009.02139.x]

[65]   Validov, S.Z.; Kamilova, F.D.; Lugtenberg, B.J.J. Monitoring of pathogenic and non-pathogenic *Fusarium oxysporum* strains during tomato plant infection. *Microb. Biotechnol.,* **2011**, *4*(1), 82-88.
[http://dx.doi.org/10.1111/j.1751-7915.2010.00214.x] [PMID: 21255375]

[66]   Gan, Z.; Marquardt, R.R.; Abramson, D.; Clear, R.M. The characterization of chicken antibodies raised against *Fusarium spp.* by enzyme-linked immunosorbent assay and immunoblotting. *Int. J. Food Microbiol.,* **1997**, *38*(2-3), 191-200.
[http://dx.doi.org/10.1016/S0168-1605(97)00108-6] [PMID: 9506284]

[67]   Vives, M.C.; Velázquez, K.; Pina, J.A.; Moreno, P.; Guerri, J.; Navarro, L. Identification of a new enamovirus associated with citrus vein enation disease by deep sequencing of small RNAs. *Phytopathology,* **2013**, *103*(10), 1077-1086.
[http://dx.doi.org/10.1094/PHYTO-03-13-0068-R] [PMID: 23718835]

[68]   Koonin, E.V. The phylogeny of RNA-dependent RNA polymerases of positive-strand RNA viruses. *J. Gen. Virol.,* **1991**, *72*(9), 2197-2206.
[http://dx.doi.org/10.1099/0022-1317-72-9-2197] [PMID: 1895057]

[69]   Olmos, A.; Bertolini, E.; Cambra, M. Simultaneous and co-operational amplification (Co-PCR): A new concept for detection of plant viruses. *J. Virol. Methods,* **2002**, *106*(1), 51-59.
[http://dx.doi.org/10.1016/S0166-0934(02)00132-5] [PMID: 12367729]

[70]   He, Y.; Yang, Z.; Hong, N.; Wang, G.; Ning, G.; Xu, W. Deep sequencing reveals a novel closterovirus associated with wild rose leaf rosette disease. *Mol. Plant Pathol.,* **2015**, *16*(5), 449-458.
[http://dx.doi.org/10.1111/mpp.12202] [PMID: 25187347]

[71]   Hayashi, K. PCR-SSCP: A simple and sensitive method for detection of mutations in the genomic DNA. *Genome Res.,* **1991**, *1*(1), 34-38.
[http://dx.doi.org/10.1101/gr.1.1.34] [PMID: 1842918]

[72]   Kitagawa, T.; Sakamoto, Y.; Furumi, K.; Ogura, H. Novel enzyme immunoassays for specific detection of *Fusarium oxysporum* f. sp. *cucumerinum* and for general detection of various *Fusarium* species. *Phytopathology,* **1989**, *79*(2), 162-165.
[http://dx.doi.org/10.1094/Phyto-79-162]

[73]   Brill, L.M.; McClary, R.D.; Sinclair, J.B. Analysis of two ELISA formats and antigen preparations using polyclonal antibodies against *Phomopsis longicolla*. *Phytopathology,* **1994**, *84*(2), 173-179.
[http://dx.doi.org/10.1094/Phyto-84-173]

[74]   Buermans, H.P.J.; den Dunnen, J.T. Next generation sequencing technology: Advances and applications. *Biochim. Biophys. Acta Mol. Basis Dis.,* **2014**, *1842*(10), 1932-1941.
[http://dx.doi.org/10.1016/j.bbadis.2014.06.015] [PMID: 24995601]

[75]   Kwon, J.Y.; Hong, J.S.; Kim, M.J.; Choi, S.H.; Min, B.E.; Song, E.G.; Kim, H.H.; Ryu, K.H. Simultaneous multiplex PCR detection of seven cucurbit-infecting viruses. *J. Virol. Methods,* **2014**, *206*, 133-139.
[http://dx.doi.org/10.1016/j.jviromet.2014.06.009] [PMID: 24937806]

[76]   Gambino, G. Multiplex RT-PCR method for the simultaneous detection of nine grapevine viruses. In: *in Plant Virology Protocols: New Approaches to Detect Viruses and Host Responses,* 3rd; Uyeda, I.; Masuta, C., Eds.; Humana Press: New York, NY, **2015**; pp. 39-47.
[http://dx.doi.org/10.1007/978-1-4939-1743-3_4]

[77]   Zhao, X.; Liu, X.; Ge, B.; Li, M.; Hong, B. A multiplex RT-PCR for simultaneous detection and identification of five viruses and two viroids infecting chrysanthemum. *Arch. Virol.,* **2015**, *160*(5), 1145-1152.
[http://dx.doi.org/10.1007/s00705-015-2360-z] [PMID: 25698104]

[78]   Wu, Q.; Ding, S.W.; Zhang, Y.; Zhu, S. Identification of viruses and viroids by next-generation sequencing and homology-dependent and homology-independent algorithms. *Annu. Rev. Phytopathol.,* **2015**, *53*(1), 425-444.
[http://dx.doi.org/10.1146/annurev-phyto-080614-120030] [PMID: 26047558]

[79]   Jones, S.; Baizan-Edge, A.; MacFarlane, S.; Torrance, L. Viral diagnostics in plants using next generation sequencing: Computational analysis in practice. *Front. Plant Sci.,* **2017**, *8*, 1770.
[http://dx.doi.org/10.3389/fpls.2017.01770] [PMID: 29123534]

[80]   Rott, M.; Xiang, Y.; Boyes, I.; Belton, M.; Saeed, H.; Kesanakurti, P.; Hayes, S.; Lawrence, T.; Birch, C.; Bhagwat, B.; Rast, H. Application of next generation sequencing for diagnostic testing of tree fruit viruses and viroids. *Plant Dis.,* **2017**, *101*(8), 1489-1499.

[http://dx.doi.org/10.1094/PDIS-03-17-0306-RE] [PMID: 30678581]

[81]   Villamor, D.E.V.; Mekuria, T.A.; Pillai, S.S.; Eastwell, K.C. High-throughput sequencing identifies novel viruses in nectarine: Insights to the etiology of stem pitting disease. *Phytopathology,* **2016**, *106*(5), 519-527.
[http://dx.doi.org/10.1094/PHYTO-07-15-0168-R] [PMID: 26780433]

[82]   Maree, H.J.; Fox, A.; Al Rwanih, M.; Boonham, N.; Candresse, T. Application of HTS for routine plant virus diagnostics: State of the art and challenges. *Front. Plant Sci.,* **2018**, *9*, 1082.
[http://dx.doi.org/10.3389/fpls.2018.01082] [PMID: 30210506]

[83]   Cheng, Y.; Tang, X.; Gao, C.; Li, Z.; Chen, J.; Guo, L.; Wang, T.; Xu, J. Molecular diagnostics and pathogenesis of fungal pathogens on bast fiber crops. *Pathogens,* **2020**, *9*(3), 223.
[http://dx.doi.org/10.3390/pathogens9030223] [PMID: 32197350]

[84]   Rajesh, T.; Jaya, M. Next-generation sequencing methods. In: *in Current Developments in Biotechnology and Bioengineering*; Gunasekaran, P.; Noronha, S.; Pandey, A., Eds.; Elsevier: Amsterdam, **2017**; pp. 143-158.
[http://dx.doi.org/10.1016/B978-0-444-63667-6.00007-9]

[85]   Moya, A.; Holmes, E.C.; González-Candelas, F. The population genetics and evolutionary epidemiology of RNA viruses. *Nat. Rev. Microbiol.,* **2004**, *2*(4), 279-288.
[http://dx.doi.org/10.1038/nrmicro863] [PMID: 15031727]

[86]   Kreuze, J.F.; Perez, A.; Untiveros, M.; Quispe, D.; Fuentes, S.; Barker, I.; Simon, R. Complete viral genome sequence and discovery of novel viruses by deep sequencing of small RNAs: A generic method for diagnosis, discovery and sequencing of viruses. *Virology,* **2009**, *388*(1), 1-7.
[http://dx.doi.org/10.1016/j.virol.2009.03.024] [PMID: 19394993]

[87]   Verdin, E.; Wipf-Scheibel, C.; Gognalons, P.; Aller, F.; Jacquemond, M.; Tepfer, M. Sequencing viral siRNAs to identify previously undescribed viruses and viroids in a panel of ornamental plant samples structured as a matrix of pools. *Virus Res.,* **2017**, *241*, 19-28.
[http://dx.doi.org/10.1016/j.virusres.2017.05.019] [PMID: 28576697]

[88]   Withers, S.; Gongora-Castillo, E.; Gent, D.; Thomas, A.; Ojiambo, P.S.; Quesada-Ocampo, L.M. Using next-generation sequencing to develop molecular diagnostics for *Pseudoperonospora cubensis*, the cucurbit downy mildew pathogen. *Phytopathology,* **2016**, *106*(10), 1105-1116.
[http://dx.doi.org/10.1094/PHYTO-10-15-0260-FI] [PMID: 27314624]

[89]   Martin, R.R.; Constable, F.; Tzanetakis, I.E. Quarantine regulations and the impact of modern detection methods. *Annu. Rev. Phytopathol.,* **2016**, *54*(1), 189-205.
[http://dx.doi.org/10.1146/annurev-phyto-080615-100105] [PMID: 27491434]

[90]   Mumford, R.A.; Macarthur, R.; Boonham, N. The role and challenges of new diagnostic technology in plant biosecurity. *Food Secur.,* **2016**, *8*(1), 103-109.
[http://dx.doi.org/10.1007/s12571-015-0533-y]

[91]   Cuppels, D.A.; Elmhirst, J. Disease development and changes in the natural *Pseudomonas syringae pv.* tomato populations on field tomato plants. *Plant Dis.,* **1999**, *83*(8), 759-764.
[http://dx.doi.org/10.1094/PDIS.1999.83.8.759] [PMID: 30845564]

[92]   Lawton, M.B.; MacNeill, B.H. Occurrence of race 1 of *Pseudomonas syringae pv.* tomato on field tomato in south-western Ontario. *Can. J. Plant Pathol.,* **1986**, *8*(1), 85-88.
[http://dx.doi.org/10.1080/07060668609501847]

[93]   Javed, B; Nawaz, K; Munazir, M phytochemical analysis and antibacterial activity of tannins extracted from *salix alba L* . Against different gram-positive and gram-negative bacterial strainsIran. *J. Sci. Technol. Trans. A Sci.,* **2020**, *44*, 1-12.

[94]   Rai, M.; Ingle, A.P.; Pandit, R.; Paralikar, P.; Gupta, I.; Chaud, M.V.; dos Santos, C.A. Broadening the spectrum of small-molecule antibacterials by metallic nanoparticles to overcome microbial resistance. *Int. J. Pharm.,* **2017**, *532*(1), 139-148.

[http://dx.doi.org/10.1016/j.ijpharm.2017.08.127] [PMID: 28870767]

[95]   Pareek, V.; Gupta, R.; Panwar, J. Do physico-chemical properties of silver nanoparticles decide their interaction with biological media and bactericidal action? A review. *Mater. Sci. Eng. C*, **2018**, *90*, 739-749.
[http://dx.doi.org/10.1016/j.msec.2018.04.093] [PMID: 29853145]

[96]   Rajeshkumar, S.; Bharath, L.V.; Geetha, R. *Broad spectrum antibacterial silver nanoparticle green synthesis: characterization, and mechanism of actionGreen Synthesis, Characterization and Applications of Nanoparticles*; Elsevier: Amsterdam, **2019**, pp. 429-444.
[http://dx.doi.org/10.1016/B978-0-08-102579-6.00018-6]

[97]   Lucas, J.A. Plant immunisation: from myth to SAR. *Pestic. Sci.,* **1999**, *55*(2), 193-196.
[http://dx.doi.org/10.1002/(SICI)1096-9063(199902)55:2<193::AID-PS883>3.0.CO;2-5]

[98]   Cohen, Y.; Niderman, T.; Mosinger, E.; Fluhr, R. Beta-aminobutyric acid induces the accumulation of pathogenesisrelated proteins in tomato *(Lycopersicon esculentum L.)* plants and resistance to late blight infection caused by Phytophthora in control of wheat diseases by a benzothiadiazole-derivative and modern fungicides. *J. Plant Dis. Prot.,* **1994**, *106*, 466-475.

[99]   Benhamou, N.; Bélanger, R.R. Benzothiadiazole-mediated induced resistance to *Fusarium oxysporum* f. sp. *radicis-lycopersici* in Tomato. *Plant Physiol.,* **1998**, *118*(4), 1203-1212.
[http://dx.doi.org/10.1104/pp.118.4.1203] [PMID: 9847094]

[100]   Javed, B.; Mashwani, Z.R. Synergistic effects of physicochemical parameters on bio-fabrication of mint silver nanoparticles: structural evaluation and action against HCT116 colon cancer cellsInt. *Int. J. Nanomedicine,* **2020**, *15*, 3621-3637.
[http://dx.doi.org/10.2147/IJN.S254402] [PMID: 32547018]

[101]   Deresinski, S. Bacteriophage therapy: Exploiting smaller fleas. *Clin. Infect. Dis.,* **2009**, *48*(8), 1096-1101.
[http://dx.doi.org/10.1086/597405] [PMID: 19275495]

[102]   Rajesh, R.W.; Rahul, M.S.; Ambalal, N.S. Trichoderma: A significant fungus for agriculture and environment. *Afr. J. Agric. Res.,* **2016**, *11*(22), 1952-1965.
[http://dx.doi.org/10.5897/AJAR2015.10584]

[103]   Wamonje, F.O.; Donnelly, R.; Tungadi, T.D.; Murphy, A.M.; Pate, A.E.; Woodcock, C.; Caulfield, J.; Mutuku, J.M.; Bruce, T.J.A.; Gilligan, C.A.; Pickett, J.A.; Carr, J.P. Different plant viruses induce changes in feeding behavior of specialist and generalist aphids on common bean that are likely to enhance virus transmission. *Front. Plant Sci.,* **2020**, *10*, 1811.
[http://dx.doi.org/10.3389/fpls.2019.01811] [PMID: 32082355]

[104]   Cilliers, M.; van Wyk, S.G.; van Heerden, P.D.R.; Kunert, K.J.; Vorster, B.J. Identification and changes of the drought-induced cysteine protease transcriptome in soybean *(Glycine max)* root nodules. *Environ. Exp. Bot.,* **2018**, *148*, 59-69.
[http://dx.doi.org/10.1016/j.envexpbot.2017.12.005]

[105]   Scholthof, K.B.G.; Adkins, S.; Czosnek, H.; Palukaitis, P.; Jacquot, E.; Hohn, T.; Hohn, B.; Saunders, K.; Candresse, T.; Ahlquist, P.; Hemenway, C.; Foster, G.D. Top 10 plant viruses in molecular plant pathology. *Mol. Plant Pathol.,* **2011**, *12*(9), 938-954.
[http://dx.doi.org/10.1111/j.1364-3703.2011.00752.x] [PMID: 22017770]

[106]   Lockhart, B.E.L. Evidence for a circular double-stranded DNA genome in a second group of plant viruses. *Phytopathology,* **1990**, *80*, 127-131.
[http://dx.doi.org/10.1094/Phyto-80-127]

[107]   Yang, L.J.; Hidaka, M.; Masaki, H.; Uozumi, T. Detection of potato virus Y P1 protein in infected cells and analysis of its cleavage site. *Biosci. Biotechnol. Biochem.,* **1998**, *62*(2), 380-382.
[http://dx.doi.org/10.1271/bbb.62.380] [PMID: 9532800]

[108]   Hinrichs-Berger, J.; Berger, S.; Buchenauer, B. The P1 protein of Potato virus Y is transiently

accumulated in systemically infected leaves of tobacco plants. *J. Plant Dis. Prot.,* **2003**, *110*, 568-571.

[109]   Domingo-Calap, M.L.; Chase, O.; Estapé, M.; Moreno, A.B.; López-Moya, J.J. The P1 Protein of *Watermelon mosaic virus* compromises the activity as rna silencing suppressor of the P25 Protein of *Cucurbit yellow stunting disorder virus. Front. Microbiol.,* **2021**, *12*, 645530.
[http://dx.doi.org/10.3389/fmicb.2021.645530] [PMID: 33828542]

[110]   Coleman, J.J. The *F usarium solani* species complex: Ubiquitous pathogens of agricultural importance. *Mol. Plant Pathol.,* **2016**, *17*(2), 146-158.
[http://dx.doi.org/10.1111/mpp.12289] [PMID: 26531837]

[111]   Nierman, W.C.; Pain, A.; Anderson, M.J.; Wortman, J.R.; Kim, H.S.; Arroyo, J.; Berriman, M.; Abe, K.; Archer, D.B.; Bermejo, C.; Bennett, J.; Bowyer, P.; Chen, D.; Collins, M.; Coulsen, R.; Davies, R.; Dyer, P.S.; Farman, M.; Fedorova, N.; Fedorova, N.; Feldblyum, T.V.; Fischer, R.; Fosker, N.; Fraser, A.; García, J.L.; García, M.J.; Goble, A.; Goldman, G.H.; Gomi, K.; Griffith-Jones, S.; Gwilliam, R.; Haas, B.; Haas, H.; Harris, D.; Horiuchi, H.; Huang, J.; Humphray, S.; Jiménez, J.; Keller, N.; Khouri, H.; Kitamoto, K.; Kobayashi, T.; Konzack, S.; Kulkarni, R.; Kumagai, T.; Lafton, A.; Latgé, J.P.; Li, W.; Lord, A.; Lu, C.; Majoros, W.H.; May, G.S.; Miller, B.L.; Mohamoud, Y.; Molina, M.; Monod, M.; Mouyna, I.; Mulligan, S.; Murphy, L.; O'Neil, S.; Paulsen, I.; Peñalva, M.A.; Pertea, M.; Price, C.; Pritchard, B.L.; Quail, M.A.; Rabbinowitsch, E.; Rawlins, N.; Rajandream, M.A.; Reichard, U.; Renauld, H.; Robson, G.D.; de Córdoba, S.R.; Rodríguez-Peña, J.M.; Ronning, C.M.; Rutter, S.; Salzberg, S.L.; Sanchez, M.; Sánchez-Ferrero, J.C.; Saunders, D.; Seeger, K.; Squares, R.; Squares, S.; Takeuchi, M.; Tekaia, F.; Turner, G.; de Aldana, C.R.V.; Weidman, J.; White, O.; Woodward, J.; Yu, J.H.; Fraser, C.; Galagan, J.E.; Asai, K.; Machida, M.; Hall, N.; Barrell, B.; Denning, D.W. Genomic sequence of the pathogenic and allergenic filamentous fungus Aspergillus fumigatus. *Nature,* **2005**, *438*(7071), 1151-1156.
[http://dx.doi.org/10.1038/nature04332] [PMID: 16372009]

[112]   Zhang, Y.; Yang, H.; Turra, D.; Zhou, S.; Ayhan, D.H.; DeIulio, G.A.; Guo, L.; Broz, K.; Wiederhold, N.; Coleman, J.J.; Donnell, K.O.; Youngster, I.; McAdam, A.J.; Savinov, S.; Shea, T.; Young, S.; Zeng, Q.; Rep, M.; Pearlman, E.; Schwartz, D.C.; Di Pietro, A.; Kistler, H.C.; Ma, L.J. The genome of opportunistic fungal pathogen Fusarium oxysporum carries a unique set of lineage-specific chromosomes. *Commun. Biol.,* **2020**, *3*(1), 50.
[http://dx.doi.org/10.1038/s42003-020-0770-2] [PMID: 32005944]

[113]   Wycoff, K.L.; Jellison, J.; Ayers, A.R. Monoclonal antibodies to glycoprotein antigens of a fungal plant pathogen, Phytophthora *megasperma f.sp.* glycinea. *Plant Physiol.,* **1987**, *85*(2), 508-515.
[http://dx.doi.org/10.1104/pp.85.2.508] [PMID: 16665728]

[114]   De Bernardis, F.; Molinari, A.; Boccanera, M.; Stringaro, A.; Robert, R.; Senet, J.M.; Arancia, G.; Cassone, A. Modulation of cell surface-associated mannoprotein antigen expression in experimental candidal vaginitis. *Infect. Immun.,* **1994**, *62*(2), 509-519.
[http://dx.doi.org/10.1128/iai.62.2.509-519.1994] [PMID: 7507895]

[115]   Hitchcock, P.; Gray, T.R.G.; Frankland, J.C. Production of a monoclonal antibody specific to *Mycena galopus* mycelium. *Mycol. Res.,* **1997**, *101*(9), 1051-1059.
[http://dx.doi.org/10.1017/S0953756297003602]

[116]   Bridge, P.D.; Couteaudier, Y.; Clarkson, J.M. *Molecular variability of fungal pathogens,* 4th; CAB International: Wallingford, UK, **1998**.
[http://dx.doi.org/10.1079/9780851992662.0000]

[117]   Paris, S.; Debeaupuis, J.P.; Crameri, R.; Carey, M.; Charlès, F.; Prévost, M.C.; Schmitt, C.; Philippe, B.; Latgé, J.P. Conidial hydrophobins of *Aspergillus fumigatus. Appl. Environ. Microbiol.,* **2003**, *69*(3), 1581-1588.
[http://dx.doi.org/10.1128/AEM.69.3.1581-1588.2003] [PMID: 12620846]

[118]   Latgé, J.P.; Mouyna, I.; Tekaia, F.; Beauvais, A.; Debeaupuis, J.P.; Nierman, W. Specific molecular features in the organization and biosynthesis of the cell wall of *Aspergillus fumigatus. Med. Mycol.,* **2005**, *43*(s1) 1, 15-22.

[http://dx.doi.org/10.1080/13693780400029155] [PMID: 16110787]

[119] Bruneau, J.M.; Magnin, T.; Tagat, E.; Legrand, R.; Bernard, M.; Diaquin, M.; Fudali, C.; Latgé, J.P. Proteome analysis of *Aspergillus fumigatus* identifies glycosylphosphatidylinositol-anchored proteins associated to the cell wall biosynthesis. *Electrophoresis,* **2001**, *22*(13), 2812-2823.
[http://dx.doi.org/10.1002/1522-2683(200108)22:13<2812::AID-ELPS2812>3.0.CO;2-Q]   [PMID: 11545413]

[120] Arruda, M.P.; Brown, P.; Brown-Guedira, G.; Krill, A.M.; Thurber, C.; Merrill, K.R.; Foresman, B.J.; Kolb, F.L. Genome-wide association mapping of Fusarium head blight resistance in wheat using genotyping-by-sequencing. *Plant Genome,* **2016**, *9*(1), 22-33.
[http://dx.doi.org/10.3835/plantgenome2015.04.0028] [PMID: 27898754]

[121] Manikandan, R.; Harish, S.; Karthikeyan, G.; Raguchander, T. Differentially expressed proteins responsible for virulence on tomato plants. *Front. Microbiol.,* **2018**, *9*, 420.
[http://dx.doi.org/10.3389/fmicb.2018.00420] [PMID: 29559969]

[122] Ramarathnam, R.; Bo, S.; Chen, Y.; Fernando, W.G.D.; Xuewen, G.; de Kievit, T. Molecular and biochemical detection of fengycin- and bacillomycin D-producing *Bacillus* spp., antagonistic to fungal pathogens of canola and wheat. *Can. J. Microbiol.,* **2007**, *53*(7), 901-911.
[http://dx.doi.org/10.1139/W07-049] [PMID: 17898845]

[123] Pineda, M.; Pérez-Bueno, M.L.; Barón, M. Detection of bacterial infection in melon plants by classification methods based on imaging data. *Front. Plant Sci.,* **2018**, *9*, 164.
[http://dx.doi.org/10.3389/fpls.2018.00164] [PMID: 29491881]

CHAPTER 4

Viral Diseases of Legumes and Their Managements

Pirtunia Nyadzani Mushadu[1,*]

[1] *Department of Biodiversity, School of Molecular and Life Sciences, Faculty of Science and Agriculture, University of Limpopo, Limpopo Province, Republic of South Africa*

Abstract: Legumes are very important food crops that are widely cultivated for their high-quality proteins, oils, and vitamins throughout the world. In total, 168 viruses are officially assigned by the International Committee on Taxonomy of Viruses. These viruses belong to 39 genera in 16 families and have been reported to affect various leguminous crops in different parts of the world. Among these viruses, TSWV (Tomato Spotted Wilt Virus), CMV (Cucumber Mosaic Virus), TMV (Tobacco Mosaic Virus), BYMV (Bean yellow mosaic virus), and BCMV (Bean Common Mosaic Virus) have global economic importance. This review therefore focuses on the economic importance of the abovementioned viruses influencing legume growth and development by looking at aspects such as viral traits, transmission, viral biology, plant host symptoms and the options used to control and manage some viruses such as the CMV (Cucumber Mosaic Virus), TMV (Tobacco Mosaic Virus), BYMV (Bean yellow mosaic virus), and BCMV (Bean Common Mosaic Virus).

Keywords: Crop yield, Disease management, Legumes, Plant virus, Viral morphology, Infection biology.

INTRODUCTION

Legumes belong to the family Leguminosae (Fabaceae), which is regarded as one of the largest and most important families of flowering plants, constituting about 650 to 750 genera, 18,000 to 19,000 species of herbs, climbers, shrubs, and trees [1]. The family is regarded as podded fruits, and the commonly used legumes include peas, lentils, peanuts, cowpeas, chickpeas, clovers, kidneys, mung beans, pigeon peas, soybean, and vetches. Legumes are used as human and animal food since they are the richest source of protein, starch, minerals, and vitamins. They also play an important role in agriculture and agroforestry by improving soil quality. These legumes can convert atmospheric nitrogen into nitrogenous compounds that are usable by plants [2]. The main limiting factor in legume

---

* **Corresponding author Pirtunia Nyadzani Mushadu:** Department of Biodiversity, School of Molecular and Life Sciences, Faculty of Science and Agriculture, University of Limpopo, Limpopo Province, Republic of South Africa; Tel: +2771-019-8740; E-mail: nyadzanimushadu@gmail.com

Phetole Mangena & Sifau A. Adejumo (Eds.)

production is yield losses that are due to pests and disease. Among all various pathogens, viruses are considered to pose the most significant effect [3]. The cultivated legumes are susceptible to natural infections caused by viruses, where resultant diseases cause a severe impact on the vegetative growth and productivity of legumes worldwide.

In total, 168 viruses belonging to 39 genera and 16 families were recorded in different parts of the world by the International Committee on Taxonomy of Viruses and they were attributed to major losses in various leguminous crops [4]. TSWV (Tomato Spotted Wilt Virus), CMV (Cucumber Mosaic Virus), TMV (Tobacco Mosaic Virus), BYMV (Bean yellow mosaic virus), and BCMV (Bean Common Mosaic Virus) are the most economically important viruses responsible for legume losses worldwide [5]. Among these, three viruses (TSWV, CMV, and TMV) are among the top economically important plant viruses mostly in non-leguminous crops [6], meanwhile, two viruses (BCMV and BYMV) are particularly regarded as economically important in reducing legume growth and productivity. Most of the legume-infecting viruses are seed-borne, with viral transmission vectored through insect pests [5]. Therefore, this chapter discusses the biology and management of some of the economically important plant viruses reported to date and evaluates their influence on the growth and productivity of leguminous crops.

TOBACCO MOSAIC VIRUS

The tobacco mosaic virus was discovered for the first time in 1879 by Adolf Mayer in the Netherlands [7]. In plant virology, the TMV is the most ancient virus and a member of the *Tobamovirus* group, which includes the *Odontoglossum* ringspot virus as well as the Sammon's opuntia virus. Numerous strains of Tobacco Mosaic Virus exist, where each strain causes different symptoms in both fruit and foliage crops [8, 9]. However, TMV can remain infective for many years, while attached to the materials used for plant support such as soil particles, culture medium, greenhouse surface, and greenhouse structures [9]. This is due to its ability to withstand high temperatures of up to 50°C [8, 10, 11]. TMV is the positive sense single-stranded (ssRNA) virus [8], however, it was also recognized first because of its easiness to affect plants and noticeable symptoms [12].

Tobacco mosaic virus was reported to be a widely distributed virus that affects several vegetables, ornamental and leguminous plants, as well as various species in Solanaceae [12, 13]. The TMV is not transmitted by insects, nematodes, or other vectors, however, it has been reported to be transmitted easily by virus-infested saps [10], particularly, through direct contact with wounded areas on the surfaces of plants [11, 14]. This virus can also be transferred by grafting seed coats to new plants from the infected mother plants [12]. The virus can also be disseminated mechanically during normal field operations and human activities [9, 10].

Infection Biology

Different pathogens that infect plants interfere with various physiological functions which often results in the development of different symptoms. Changes that result from the multiplication of the virus cause a reduction in plant yield and reduced quality of the product [15]. Symptoms in infected plants vary according to the strains of the Tobacco Mosaic Virus, the type of plant species that get infected, and the developmental stage at which a particular plant is infected [12]. Pepper plants infected by the virus developed recognisable disease symptoms early in their seedling and immature stages [15]. However, all viral diseases are generally associated with direct or indirect biochemical aberration induced by the virus. It has also been reported that the first symptoms after virus infection appear as necrosis and chlorosis on the uppermost younger leaves along the main veins, followed by wilting and leaf spots [15].

Generally, the infected pepper shows reductions in leaf numbers, leaf area, mottled leaves, deformed and distorted leaf phenotype, as well as stunted shoot growth. As such, this contributes to a reduction in photosynthetic activity. Photosynthetic activities provide the plant with the energy that is required for its growth and defence against diseases and pests. Since MTV is associated with the reduction in leaf numbers and total leaf area, which are parameters linked to photosynthesis, this reduction causes a decline in plant growth, resulting in shortened slender plant stems, and reduced biomass. Infections caused by TMV also cause a reduction in relative water content and photosynthetic pigments (chlorophylls) [15]. Compared to leguminous crops, infected tobacco plants produce TMV more abundantly than enclosed crystallized virion bodies [11]. The first symptoms in tobacco plants are vein clearing at the youngest leaves, followed by a distinct mosaic of light-green and dark-green areas at early developmental stages [10]. Mosaic symptom development involves changes in chloroplast structures whereby some of the TMVs are detected earlier in chloroplast metabolism [16]. The virus causes light and dark green mottled areas in tomato leaves. In most cases, the area appearing dark green becomes thicker than the portion of the leaf which is lighter in colour. Young shoot growth usually becomes stunted with distorted leaf curling. Additionally, some strains produce mottling, streaking and death of the fruits [12]. Tobacco Mosaic Virus infection resembles water stress which also causes an increase in cytoplasmic ABA. Generally, the TMV infection causes a two to six-fold increase in the concentration of ABA in the leaf. The ABA is important in controlling the inhibition of growth that results from the negative effects of these infectious pathogens [16].

Disease Management

According to Henn [12], some researchers suggested that the TMV may be managed and controlled by disinfecting the tomato, tobacco, and other crop seeds with trisodium phosphate or calcium hypochlorite, respectively [9]. The virus can also be prevented by removing the soil that has been used to grow infected plants since it contains infected species in roots, then leaves and other plant debris [12]. Household bleach of 10% solution can effectively kill TMV. It can be used for 1 minute to disinfect benches, equipment and surface areas that were in contact with the virus or infected plant materials [9, 12]. All infected plants should be placed in bags and removed from the all farming operations, as well as be destroyed to help in managing the virus. The weeds that serve as TMV hosts must be controlled in and around the planting beds [10]. Virus-free transplants should be used to control TMV incidences. Before the first cultivation, any plants that show symptoms of the mosaic virus should be removed. Workers should wash their hands and forearms with soap and water before entering the field to avoid the introduction of the virus [9, 10].

Whole milk or 20% of powdered non-fat skim milk with a surfactant of between 0,1% to 20% is slightly more effective than the detergent soap when removing the virus from the hands or elbows [10]. The application of pesticides can reduce the virus' diseases by minimizing the vector population. Plant viruses can also be managed by using natural extracts from plants, animals, microorganisms, and other commercial bio-pesticides. Polysaccharides from *Arctium lappa* root extracts are reported to be effective against TMV. When compared with control treatments, it amplifies the level of transcription of multiple defence-related proteins (DRPs) and enzymes within 24 hours after inoculation with DRPs that suppress and effectively inhibit the replication of the virus [8].

BEAN YELLOW MOSAIC VIRUS

Bean yellow mosaic virus (BYMV) belongs to the *Potyvirus* genus and family Potyviridae. Several reports show BMYV as an important plant virus economically, as it infects many crops from the Fabaceae family and other members of Liliaceae [17 - 19]. Several species within the families of monocotyledons and dicotyledons can be infected by BMYV, which is known to happen worldwide [19]. The transmission of the Bean Yellow Mosaic Virus occurs non-persistently in several aphid's species and through seeds in broad beans [20, 21]. The virus can also be transmitted to healthy bean plants through mechanical inoculation and pollen [19, 22].

Infection Biology

Once the virus infects the plants, its symptoms appear as dark and yellowish-green areas. Blight yellow spots in infected plants intensify in colour as the plants get older [20]. BYMV-infected plants show leaf distortion, mosaic vein clearing and stunting [20]. The leaves are rolled up and this leads to a reduction in photosynthetic rate which reduces plant growth and delays plant maturity (bean package). Pods of the infected legume plants may be malformed, fewer, distorted and have an uneven surface which shows light green mottle [19, 22]. In a plant, a virus infection may occur or show its development before flowering at any time. It was noted that in faba bean seed, BYMV produced necrotic rings and discoloration [23], while in the leaf cell of broad and common beans (*Vicia faba* L and *Phoseolus vulgaris* L.), BYMV produced crystalline inclusions which are abundantly found in a form of crystals in the cytoplasm, nucleolus and nucleus when viewed under a microscope [24].

Disease Management

According to Kumari and Makkouk [25], the virus infection in crops cannot be cured in any direct practical way, hence all control strategies currently emphasize measures that prevent or reduce the virus infection. BYMV control measures are classified into, those measures that are directed towards vector avoidance/ incidence reduction, the ones that control the virus and those that are integrated by combining all the possible components of control in a way that it can be applied as one control package in the farm or field [25].

Measures Targeting Sources of Infection

Since more than 50% of the virus affecting faba bean is seed-borne, it is always recommended that farmers use seeds that are virus-free for planting. The above-mentioned technique can also be used in cases where the virus is transmitted by active vectors. Phytosanitary control, such as roguing or the removal of asymptomatic plants can also be used to remove virus infection sources from the field that is affected. It was reported that practising roguing of virus-infected crops 2-3 times during the early growing season efficiently minimized the incidence of primary infection in Egypt. Additionally, the spatial isolation technique, which involves avoiding over wintering and over a summering crop that serves as a source of infection, works more effectively for non-persistent viruses such as BYMV [25].

Method Directed Towards Vector Avoidance/Incidence Reduction

Cultural Practices

It has been reported that virus incidences in crops can be effectively reduced by following cultural practices such as the date of planting, narrow row spacing, high rate of seedling and using cultivars that mature early. Planting date manipulation can be used as a standard virus control measure as it helps in avoiding the exposure of young plants to peak populations of vector since they are more vulnerable in the early stage of growth. In Syria and Egypt, the faba bean crops that were planted early in September got attacked severely, leading to 100% BYMV infection. However, when farmers delayed sowing until October/ November, the faba bean crops were less in quantity which resulted in lower virus infection [25, 26].

Virus Vector Control

The spread of BMYV vectored by insects can be reduced by applying insecticides. This method is often ineffective due to its incapability to kill the vector carrying non-persistence virus fast enough to prevent probing and subsequent virus inoculation to the sprayed plant. Newer generation of synthetic pyrethroids is the most effective type of insecticides for controlling non-persistently transmitted aphid-borne viruses. The use of systemic seed treatment insecticide Imidacoprid at a rate of 0.5-28 g a.i./kg seed in the field experiment of ICARDA gave significant protection against BYMV infection [25].

Breeding for Vector Resistance

Over 1000 lines of resistance of faba bean to aphid have been screened and from the list, 36 have been classified as being resistant to Bean Yellow Mosaic Virus. The development of insects that are resistant to pesticides results in the increase in the interest of breeding for insect's resistance in plants. In fields, the use of resistant cultivars could lead to BYMV reduction [25].

Integrated Approach

The combination of genetic resistance, cultural practices and chemical sprays could lead to improved disease control. Combining the use of resistant seeds, followed by dated planting in a particular region and the use of various insecticides, maximizes the suppression of BYDV spread, which offers farmers economical and reasonable control [26].

TOMATO SPOTTED WILT VIRUS

Tomato Spotted Wilt Virus is classified among the top 10 plant viruses by virologists, because of its scientific and economic importance worldwide [27]. TSWV belongs to the *Tospovirus* genus and it is the only plant-infecting virus in the family Bunyariridae [28, 29]. It is unique in the sense that it is in a virus class alone [30]. TSWV is an enclosed, negative-stand RNA virus that contains two glycosylated protein membranes such as a putative RNA-dependant RNA polymerase and a nucleocapsid protein [29]. The virus is also known to have a wide range of plant hosts, which can reach about 900 species in more than 70 families, with several hundred species in both monocotyledonous and dicotyledonous plants which include tomatoes and ornamentals. The tomato spotted wilt virus infects plants from families, such as Solanaceae, Asteraceae, Leguminosae, Brassicaceae, and Bromiliaceae [27 - 29].

Faba beans (*Vicia faba*) were reported as principal hosts of TSWV in other regions like Australia, Algeria, Belgium, Cyprus, and Egypt. Other legumes like chickpeas and lentils were found to be infected seriously by the virus in Brazil. Furthermore, TSWV naturally infects other legumes like soybeans, peas, mung beans and cowpeas [31]. In June 2016, there was a report on soybean infection by TSWV in Iran as well as Georgia, where the collected leaf samples of soybeanshowed viral disease symptoms such as chlorosis, mosaic, mottling and stunting. Soybeans in Korea are planted with other plants that are TSWV hosts such as pepper and tomato. Hence, the Tomato Spotted Wilt Virus infection in soybean is associated with TSWV, due to its natural occurrence which has been reported in both pepper and tomatoes [32]. TSWV is generally transmitted by nine species of strips from one plant to another. However, *Frankliniella occidentalis* is commonly called western flower strips and serves as the most efficient vector transmission that occurs in a propagative and persistent manner [28, 33].

Infection Biology

The viral disease negatively affects the physiological and biochemical aspects of a plant. Virus infection in plants leads to multiple substantial modifications at the physiological and biochemical level, which in turn disturb several vital processes of the host cell's balance and functions by using its system of protein synthesis to produce non-structural protein such as virus coat protein and nucleic acid replicating enzymes that are essential to produce a new particle of the virus [27]. Viral diseases disturb the process or the rate of photosynthesis by affecting the content of leaf pigment. When the infected plants respond to leaf pigment disruption and reduced photosynthesis process, it often triggers the production of harmful molecules called reactive oxygen species (ROS) such as hydrogen peroxide, hydroxyl radicals, oxygen, and superoxide anion in the tissue.

The ROS molecules produced as a by-product of the metabolic or photosynthesis process destroy the biological membranes and cell components which include proteins, and nucleic acid and lastly, they cause the peroxidation of lipids [27]. Lipid peroxidation, a metabolic process where ROS results in the degradation of oxidative lipids, tends to affect the cell membrane structure and its functions. Lipid peroxidation products cause damage to the DNA [34]. In the investigation of how tomato plants respond to TWVS, the reduction in leaf pigment which includes chlorophyll a, chlorophyll b, and carotenoids was observed as compared to that in a healthy tomato. The decline that was observed in the leaf pigments concentration in TSWV-infected tomatoes was attributed to the chloroplast enzyme disruption. The chloroplast enzymes play an essential role in the synthesis of leaf pigments.

It was concluded that in tomato plants, the TSWV leads to abnormal photosynthesis regulation. TSWV causes a lot of symptoms such as mosaic mottling, and yellowish leaf tissue in lettuce [27], and the first symptoms usually are seen on the young inner upper part of the leaves on one side of the plant as wilting of the margin slightly, necrotic rings spot and leaf blackening occur [35]. Those plants that are infected in an early growing season do not produce any fruits, while those ones infected after the setting of fruits produce their fruits with striking symptoms, which include chlorotic concentric ring spots, raised bumps deformation, and uneven ripening. TSWV is associated with poor-quality fruit production and low yields [30]. All the symptoms /effects depicted on virus-infected plants are always associated/connected with an inhibition in the photosynthesis pigments synthesis [27].

Tomato Spotted Wilt Virus Management

Integrated Pest Management

Most farmers that grow cucumber, tomatoes, herbs, pepper, soybean, and other legume crops integrated pest management programs, which involve the use of cultural control methods, possibly available resistant varieties, biological control agents as well as minimal use of compatible pesticides. However, the success of an IPM (integrated pest management) programme, critically depends on the choice, combination, and timing of measure application [35]. Planting disease and insect-resistance vegetable varieties, cultivation, and destruction of pest overwintering sites can reduce TSWV incidences [36].

Cultural Control

It is advisable that before planting any crops in fields and greenhouses that previously had crops infected with TSWV, proper clean-up should be done to minimize thrips carryover. The debris of previously infected crops should be removed and disposed of since they can maintain a source of strips that can invade the field, or the greenhouse and may infest new crops that are planted [35]. Cultivation operation in the fields and greenhouses should be minimized during periods when thrips are abundant because disturbances of plants at these times further encourage strips movements [37].

Rouging of TSWV-infected Plants

Plants should be carefully inspected for any possible TSWV symptoms on arrival or the first week after planting. The infected plants need to be removed from the crop as soon as they are detected. Early rouging of infected plants helps to stop the spread of the TSWV throughout the entire field. Plants that were removed must be quickly placed in a sealed bag and taken to a locally authorised disposal site. The site is generally used to prevent the plants from regrowing in the area where they are disposed. When removing the infected plants, it is recommended that a whole plant should be removed instead of a part that shows symptoms. It is recommended that weed populations that either act as possible hosts or reservoirs of the TSWV should be controlled in and around the greenhouse and in fields [35, 37].

TSWV-resistant Plant Varieties

They are limited host plant resistance to TSWV. However, currently, there are several varieties of peanuts that have been produced which showed resistance to TWSV. Some varieties of pepper, lettuce and tomato have partial resistance to TSWV. Intensive effort is being made to obtain genetically resistant varieties as this could be the most effective way of avoiding the disease [35, 37].

CUCUMBER MOSAIC VIRUS

The Cucumber Mosaic Virus was discovered for the first time in 1934, when it showed mosaic symptoms in cucumbers. CMV is the type of species that falls under *Cucumovirus* genus in the Bromoviridae family and subfamily alpha-like virus. CMV is known to have the broadest host range with approximately 1200 plant species for any plant virus in over 100 plant families [38, 39]. The host of CMV covers a wide range of vegetables, herbaceous fruits, woody or ornamentals, and weeds [40]. Cucumber Mosaic Virus is regarded as the most important and widely spread virus that is now occurring in both temperate and tropical climates worldwide. It affects many agricultural and horticultural crops as one of the most prevalent viruses. Among all the hosts, the most susceptible are listed in Cruciferae, Solanaceae, Leguminosae, Cucurbitaceae, and Compositae, however, only specific strains of the Cucumber Mosaic Virus tend to cause disease in legumes [41].

The first report of the Cucumber Mosaic Virus strain occurring naturally in dry beans (*Phaseolus vulgaris* L.) was found in the nursery plantings by Whipple and Walker. After that, it was also reported to have been noted in Iran. As described by literature, there are some strains of CMV that are known to cause seed-borne diseases in legumes, which show symptoms that are far different from those caused by BCMV (Bean Common Mosaic Virus) and any other virus. In peas, yardlong beans, dry beans, and cowpeas, CMV has been reported to be seed-borne [42]. Moreover, in some weeds, CMV is seed-borne, but it has aphid vectors and natural reservoirs which in crops account for its high incidences [43]. CMV is non-persistently transmitted by aphids *M. persicae* into healthy plants, which acquire the virus during brief probes on infected hosts [44]. Due to the complex epidemiology of the virus and its seed transmission in legumes, the CMV received special attention. The outbreak of CMV has also been reported and recorded in pepper, tobacco celery, melon, and zucchini squash [38].

Infection Biology

The particles of CMV multiply in the infected plant cell, they tend to alter biochemical compounds of cells such as chlorophyll, nucleic acids, β-carotene, and organic carbon and disrupt the physiological process that ends up affecting growth and yields. In the virus-infected plants, the biochemical changes result in a decrease in the quality and quantity of the infected crops. The symptoms and their severity are concerned with specific cellular component changes resulting from viral infections. Manifestations of external disease symptoms often cause host metabolism alteration. A decrease in total chlorophyll was reported, followed by photosynthesis reduction due to the infection of the Cucumber Mosaic Virus in *Luffa aegyptica*. The alteration of biochemical cellular constituents is reported to be directly related to the morphological deviation of plants that are infected with a virus, nevertheless, the extent of crop loss is largely determined by the visible symptoms [39, 45].

According to Li *et al.* [46], when CMV infects pepper, the common symptoms that are observed are the vein clearing, mottle, mosaic, yellow discoloration, and narrowing or shoe stringing. The symptom after infection solely depends on the age of the plant. Plants that are infected at a younger stage of development show severe symptoms while plants that are infected at a later stage of growth may be asymptomatic. At the time of infection, if the plant is young, early leaves often become pale green and slightly wrinkled. The new leaves that are emerging in a plant as it continues to grow develop a chromatic mosaic pattern that tends to enclose the whole leaf. The foliage may develop oak-leaf patterns or ringspots during growth. The pepper plants also experience severe stunting and reduction of flower formation. The fruits that the infected plants produce are small, bumpy, patchy discoloured, and malformed and tend to show depressed spots of necrotic lesions, thus resulting in reduced fruit yields and quality [46].

In peas, it was reported that the virus produces wilting and curling over the growing point which later dies. Virus-infected plants show brownish streaks along the stems and the petiole. The pods produced by the peas, remain flat and turn purplish brown in colour. The CMV strains that have been reported to infect common beans cause symptoms such as dark green vein-banding, mosaic, curling and chlorotic mottle on beans as well as in other legumes [47]. The pods of the legumes infected with the strain of CMV are mostly curled, and mottled and show reduced sizes [48].

Disease Management

There are different ways that CMV can be managed, however, due to its wide variety of alternative hosts, the eradication of the infected plants from the field plot is regarded as the main biological strategy that can be used to control CMV. The use of healthy clean and disease-free seeds also controls the cucumber virus, since it is seed-borne transmitted. A 15% trisodium phosphate solution was effectively used as a method of reducing viral disease infection by treating the seeds before they were planted in the fields or in greenhouses. Chemical insecticides and biological pesticide applications are also effective ways of controlling insect vectors by killing the vector that carries the virus before it reaches the plant. However, it has been reported that this method has limitations and disadvantages such as killing important insects and possessing environmental issues. Some insects may also become resistant to the insecticides, in such cases, the use of a developed disease-resistance cultivar is an appropriate approach that can assist in overcoming the CMV threat [46].

Control of Aphid Vector

Proper applications of the insecticide spray kill aphids before they move to other plants, thus reducing the incidence of diseases and weeds that serve as the host of CMV where the colonies that may develop should be destroyed over the large areas. Preventing the aphids from reaching plants in more sustainable control approaches, can only be achieved in nurseries [49]. Some farmers use mulches to cover the soil surface, thus delaying infection until plants are older since they can tolerate virus infection. Early sowing or late transplanting and avoiding overlapping cycles of susceptible crops in lettuce and tomato have been reported to be effective in preventing CMV diseases. In southern Italy, fruit necrosis incidences caused by CMV were reduced by coupling irregular irrigation with mulching [38, 46].

CMV-resistance Varieties

To avoid environmental issues, through the application of insecticides, CMV-resistance varieties may be used to control plant viruses. Resistance varieties of spinach, cucumber, and pepper are available, however, for the legumes, there are no good resistant cultivars that have been produced so far [49]. Some of the resistant cultivars for other crops except legumes have not been developed yet, but selected genes can be incorporated by genetic engineering into the plant genome or mild strain of CMV, which can simply be inoculated into the host plant which will be cross-protected against virus strains [38]. In the study conducted by Derbalah and EI-Sharkawy [50], the possibility of success for using nickel oxide nanostructures (NONS) as approaches for controlling CMV pathogens was demonstrated. Their study concluded that NONS is a good control for CMV infection since all tested/ investigated parameters under NONS in cucumber were good compared to non-treated plants [50].

Cultural Management

Sanitation in the field should be maintained to prevent the initial spread of the virus. Plant debris should be removed and burned after cropping season to avoid the presence of the previous virus in the field. The virus may also be eliminated by composting plants that are infected since the virus will die due to its exposure to high temperatures [51]. Leaf-colonizing *B. amyloliquefaciens* strain 5B6 capacity was evaluated to protect plants against the Cucumber Mosaic Virus for pepper in South Korea. The aerial parts of the plant were sprayed with the prepared liquids of strain 5B6. It was reported that the three years of observation of the treated field showed a consistent reduction in the CMV accumulation in plants. The beneficial effect of using 5B6 against plant viruses is that it does not affect fruit yield. In pepper plants that were treated with 5B6, there were no fruit reductions that were detected. No differences were detected between the vegetative growth (as indicated by the height of the shoot) of the pepper plants that were treated with water and 5B6. It was concluded after the observation that leaf-colonizing bacilli really does protect plants against viral diseases in the field. Nevertheless, this method has also been reported to be effective in greenhouses [52].

BEAN COMMON MOSAIC VIRUS

BCMV is regarded as the most destructive common virus that infects common beans and a range of other wild legumes that are cultivated worldwide [53, 54]. BCMV is a member of *Potyvirus* in the family of Potyviridae. It is a monoparticle, single-stranded positive-sense RNA virus [55]. BCMV has a wide range of hosts which include plants from Amaranthaceae, Chenopodiaceae, Leguminosae, Solanaceae and Tetragoniaceae families. Bean Common Mosaic Virus naturally infects wild and legume crops, such as *Phaseolus* species, *Vicia faba*, *Arachis hypogaea* (peanut) and *Vigna unguiculata* (cowpea) [54]. BCMV is transmitted in a non-persistent manner by aphids and is also seed-borne [55].

Disease Biology

Viral infection is generally associated with the alteration of physiological and biochemical processes in a particular plant. As the virus multiplies in plants, symptoms such as vein clearing, vein banding, leaf rolling and distortion, mosaic, puckering, stunned growth, purpling, mild to conspicuous yellow spots or patched, mottling and necrosis of leaves often develop [53, 56]. When the leaf rolls, it reduces the leaf area, which reduces the photosynthetic pigments and lastly the whole plant experiences decreased photosynthesis rate. Photosynthesis rate reduction in virus-infected plants is mainly due to the physical damage caused by the virus on the chloroplast structure and membrane deterioration.

Disease Management

BCMV is an aphid and seed-transmitted, hence the measure employed to control this should be based on seed and aphid control strategies in cowpeas. The incidence of BCMV was reduced by immunizing the host plants with extracts of *Clerodendrum inerme*, *Psidium guajava*, *Azadirachta indica* and *Thuja occidentalis*. Plant extracts are safe, effective and eco-friendly methods to manage BCMV diseases. The extracts offer maximum protection to the bean in both the field and greenhouse when it is applied [57].

Planting Date

The best way to control BCMV is by complying with the recommended planting date in each region. The idea is to avoid peek fights activity of the aphid's vector to minimize the time window available for the vector to colonize and infect the plant. Early planting has been reported to put the fields at greater risk for BCMV, since plants will have a higher chance of being exposed to aphids that carry the viruses. Later seedling dates tend to provide less time for aphids to feed and transmit BCMV. It has been reported that planting early in the spring provides plants with enough time to grow out of the most susceptible seedling stage before aphids move into the crop, thus this reduces the extent of the damage [58]. Worldwide researchers have developed optimum seedling planting dates for both fall and spring-planted crops [59, 60].

Cultural Control

To control BMCV incidences among beans, farmers should plant certified resistant seed cultivars. It is recommended that one should not save seeds from the previous season since they may contain the virus. The bean yield worldwide may be maximized by using resistant bean cultivars. Current and recent technologies also allow researchers to confer BCMV resistance genes such as I^2, bc-21bc-2^2, bc-1, and bc-2 to crops in the development of transgenic plants that resist BMCV infections [61].

CONCLUSION

A large number of viruses have been reported worldwide that reduce legume yields and production. However, not all of them are regarded as economically important [3]. With the increase in population, legume demand is also expected to increase rapidly since it serves as both human and animal food sources. Diseases that are caused by viruses are difficult to control and Integrated Disease Management is an effective way among other measures to control virus diseases in legumes. Single measures do not seem to eradicate or fully control the virus. Hence, to maintain and increase legume production worldwide, it is recommended that farmers should stick to Integrated approaches. It is also up to the farmers to decide which IDM to combine as they see them fit to control the virus in particular regions. Resistant varieties in some legumes have not yet been developed, as such researchers need to continue screening various legumes and viruses to develop novel, high-yielding and more resistant cultivars [35, 62].

LIST OF ABBREVIATIONS

ABA	Abscisic acid
BCMV	Bean common mosaic virus
BYMV	Bean yellow mosaic virus
CMV	Cucumber mosaic virus
DRPS	Defence related protein
DNA	Deoxyribo Nucleic acid
ICARDA	International centre for agricultural research in dry areas
IPM	Integrated pest management programme
NONS	Nitric oxide nasal spray

RNA	Ribonucleic acid
ROS	Reactive oxygen species
SSRNA	Single stranded RNA
TMV	Tobacco mosaic virus
TSWV	Tomato Spotted wilt virus

CONSENT FOR PUBLICATION

Not applicable.

CONFLICT OF INTEREST

The authors declare no conflict of interest, financial or otherwise.

ACKNOWLEDGEMENT

Declared none.

REFERENCES

[1]   Ahmed, S.; Hasan, M.M. Legumes: An overview. *RADS J Pharm Pharmacol Sci,* **2014**, *2*(1), 34-38.

[2]   Messina, M.J. Legumes and soybeans: Overview of their nutritional profiles and health effects. *Am. J. Clin. Nutr.,* **1999**, *70*(3) Suppl., 439S-450S.
        [http://dx.doi.org/10.1093/ajcn/70.3.439s] [PMID: 10479216]

[3]   Tantera, D.M. Present status of rice and legume virus diseases in Indonesia. *Trop Agric Res Ser,* **1986**, *19*, 20-32.

[4]   Chatzivassiliou, E.K. An annotated list of legume-infecting viruses in the light of metagenomics. *Plants,* **2021**, *10*(7), 1413.
        [http://dx.doi.org/10.3390/plants10071413] [PMID: 34371616]

[5]   Hema, M.; Sreenivasulu, P.; Patil, B.L.; Kumar, P.L.; Reddy, D.V.R. Tropical food legumes: Virus diseases of economic importance and their control. *Adv. Virus Res.,* **2014**, *90*, 431-505.
        [http://dx.doi.org/10.1016/B978-0-12-801246-8.00009-3] [PMID: 25410108]

[6]   Rybicki, E.P. A Top Ten list for economically important plant viruses. *Arch. Virol.,* **2015**, *160*(1), 17-20.
        [http://dx.doi.org/10.1007/s00705-014-2295-9] [PMID: 25430908]

[7]   Zaitlin, M. The discovery of the causal agent of the tobacco mosaic disease.*Discoveries in Plant Biology*; Kung, S.D.; Yang, S.F., Eds.; World Publishing: Hong Kong, **1998**, pp. 105-110.
        [http://dx.doi.org/10.1142/9789812817563_0007]

[8]   Islam, W.; Qasim, M.; Noman, A.; Tayyab, M.; Chen, S.; Wang, L. Management of tobacco mosaic virus through natural metabolites. *Rec. Nat. Prod.,* **2018**, *12*(5), 403-415.
        [http://dx.doi.org/10.25135/rnp.49.17.10.178]

[9]   Zitter, T. Tobacco Mosaic Virus control 1. Department of Plant Pathology and Plant-Microbe Biology. Cornell University College of Agriculture and Life Science. **2012**. Available from: https://cals.cornell.edu/agricultural-experiment-station (Date accessed: 04/11/2022).

[10]   McRitchie, JJ Tobacco Mosaic Virus. Florida Department of Agriculture and Consumer Services. Plant Pathology Circular No. 223. **1981**. Available from:

https://ccmedia.fdacs.gov/content/download/11230/file/pp223.pdf

[11]   Creager, A.N.H.; Scholthof, K.B.G.; Citovsky, V.; Scholthof, H.B. Tobacco mosaic virus. Pioneering research for a century. *Plant Cell,* **1999**, *11*(3), 301-308.
        [http://dx.doi.org/10.1105/tpc.11.3.301] [PMID: 10072391]

[12]   Henn, A. *Tobacco Mosaic Virus. Mississippi State University, Extension Professor, Entomology and Plant Pathology*; May 8 and June 30.

[13]   Iftikhar, Y.; Vadamalai, G.; Sajid, A.; Mubeen, M.; Ullah, M.I. Wilting of bean plants from tobacco mosaic virus from smoking tobacco in Pakistan. *Asi. J. Biolog. Lif. Sci.,* **2018**, *7*(2), 77-80.
        [http://dx.doi.org/10.5530/ajbls.2018.7.8]

[14]   Scholthof, K.B.G. Tobacco mosaic virus: A model system for plant biology. *Annu. Rev. Phytopathol.,* **2004**, *42*(1), 13-34.
        [http://dx.doi.org/10.1146/annurev.phyto.42.040803.140322] [PMID: 15283658]

[15]   Pazarlar, S.; Gümüş, M.; Öztekïn, G.B. The effects of tobacco mosaic virus infection on growth and physiological parameters in some pepper varieties (*Capsicum annuum* L.). *Not. Bot. Horti Agrobot. Cluj-Napoca,* **2013**, *41*(2), 427-433.
        [http://dx.doi.org/10.15835/nbha4129008]

[16]   Whenham, R.J.; Fraser, R.S.S.; Brown, L.P.; Payne, J.A. Tobacco-mosaic-virus-induced increase in abscisic-acid concentration in tobacco leaves. *Planta,* **1986**, *168*(4), 592-598.
        [http://dx.doi.org/10.1007/BF00392281] [PMID: 24232338]

[17]   Schulze, A.; Roberts, R.; Pietersen, G. First Report of the Detection of *Bean yellow mosaic virus* (BYMV) on *Tropaeolum majus*; *Hippeastrum* spp., and *Liatris* spp. in South Africa. *Plant Dis.,* **2017**, *101*(5), 846.
        [http://dx.doi.org/10.1094/PDIS-10-16-1446-PDN]

[18]   Mahfouze, S.A.; Khattab, E.; Gadalla, N. Resistance of faba bean accessions to bean yellow mosaic virus and broad bean stain virus. *Afr. J. Plant Sci. Biotechnol.,* **2012**, *6*(1), 60-65.

[19]   Wylie, S.J.; Coutts, B.A.; Jones, M.G.K.; Jones, R.A.C. Phylogenetic analysis of Bean yellow mosaic virus isolates from four continents: Relationship between the seven groups found and their hosts and origins. *Plant Dis.,* **2008**, *92*(12), 1596-1603.
        [http://dx.doi.org/10.1094/PDIS-92-12-1596] [PMID: 30764292]

[20]   Kaur, C.; Raj, R.; Kumar, S.; Raj, S.K. First report of *Bean yellow mosaic virus* on Cape gooseberry in India. *New Dis. Rep.,* **2014**, *29*(1), 17.
        [http://dx.doi.org/10.5197/j.2044-0588.2014.029.017]

[21]   Wang, D.; Ocenar, J.; Hamim, I.; Borth, W.B.; Fukada, M.T.; Melzer, M.J.; Suzuki, J.Y.; Wall, M.M.; Matsumoto, T.; Sun, G.F.; Ko, M.; Hu, J.S. First report of Bean yellow mosaic virus infecting nasturtium (*Tropaeolum majus*) in Hawaii. *Plant Dis.,* **2019**, *103*(1), 168.
        [http://dx.doi.org/10.1094/PDIS-06-18-1082-PDN] [PMID: 30358505]

[22]   Schwarts, H.F.; Gent, D.H.; Franc, G.D.; Harveson, R.M. Bean Yellow Mosaic. Dry Bean. High Plains IPM Guide, a cooperative effort of the University of Wyoming, University of Nebraska; Colorado State University and Montana State University, 2007.

[23]   Pardina, P.R.; Nome, C.; Reyna, P.; Muñoz, N.; Caro, E.A.; Luque, A.; Debat, H.J. Bean Yellow Mosaic Virus infecting broad bean in the green belt of Córdoba, Argentina. *BioRxiv,* **2019**, 606384.
        [http://dx.doi.org/10.1101/606384]

[24]   Weintraub, M.; Ragetli, H.W.J. Intracellular characterization of bean yellow mosaic virus-induced inclusions by differential enzyme digestion. *J. Cell Biol.,* **1968**, *38*(2), 316-328.
        [http://dx.doi.org/10.1083/jcb.38.2.316] [PMID: 4874490]

[25]   Kumari, S.G.; Makkouk, K.M. Virus diseases of faba bean (*Vicia faba* L.) in Asia and Africa. *Plant Viruses,* **2007**, *1*(1), 93-105.

[26] Jones, R.A.C.; Coutts, B.; Burchell, G.; Wylie, S. Bean yellow mosaic virus in lupins. *West Aus J of Agric,* **1997**, *38*(42), 420-424.

[27] Bondok, A.M.; Ibrahim, M.F.M. Citric and ascorbic acid drive some physiological, biochemical and molecular aspects in tomato plants inoculated with Tomato Spotted Wilt Virus (TSWV). *Mid East J Agric Res,* **2014**, *3*(4), 1248-1261.

[28] Farooq, A.A.; Akanda, A.M. Impact of tomato spotted wilt virus (TSWV) on growth contributing characters of eight tomato varieties under field condition. *Int J Sus Crop Prod,* **2007**, *2*(1), 35-44.

[29] Asano, S.; Hirayama, Y.; Matsushita, Y. Distribution of *Tomato spotted wilt virus* in dahlia plants. *Lett. Appl. Microbiol.,* **2017**, *64*(4), 297-303.
[http://dx.doi.org/10.1111/lam.12720] [PMID: 28129432]

[30] French, J.M.; Goldberg, N.P.; Randall, J.J.; Hanson, S.F. New Mexico and the southwestern US are affected by a unique population of tomato spotted wilt virus (TSWV) strains. *Arch. Virol.,* **2016**, *161*(4), 993-998.
[http://dx.doi.org/10.1007/s00705-015-2707-5] [PMID: 26721573]

[31] Zindović, J.; Bulajić, A.; Krstić, B.; Ciuffo, M.; Margaria, P.; Turina, M. First Report of *Tomato spotted wilt virus* on Pepper in Montenegro. *Plant Dis.,* **2011**, *95*(7), 882.
[http://dx.doi.org/10.1094/PDIS-03-11-0167] [PMID: 30731705]

[32] Yoon, Y.N.; Jo, Y.; Cho, W.K.; Choi, H.; Jang, Y.; Lee, Y.H.; Bae, J.Y.; Lee, B.C. First Report of *Tomato spotted wilt virus* infecting soybean in Korea. *Plant Dis.,* **2018**, *102*(2), 461.
[http://dx.doi.org/10.1094/PDIS-07-17-1051-PDN]

[33] Ogada, P.A.; Moualeu, D.P.; Poehling, H.M. Predictive models for tomato spotted wilt virus spread dynamics, considering *Frankliniella occidentalis* specific life processes as influenced by the virus. *PLoS One,* **2016**, *11*(5), e0154533.
[http://dx.doi.org/10.1371/journal.pone.0154533] [PMID: 27159134]

[34] Visilaki, A.T.; McMillan, D.C. *Lipid peroxidation. Encyclopaedia of Cancer,* 3rd; Springer-Veglag Berlin Heidelberg, **2017**, pp. 2054-2055.

[35] O'Neil, T; Bennison, J Tomato Spotted Wilt Virus in protected edible crops.*Horticultural Development Company*; Factsheet 23/10, Project Edibles, Project No. PC 289, **2010**.

[36] Mayfield, J.L.; Hudgins, J.L.; Smith, J.L. Integrated management of thrips and tomato spotted wilt virus in field-grown fresh market tomatoes. *Proc Flo State Hort Soc,* **2003**, *116*, 161-164.

[37] Kucharek, T.; Brown, L.; Johnson, F.; Funderburk, J. *Tomato spotted wilt virus of agronomic vegetable, and ornamental crops*; Plant Pathology Fact Sheet. Circular-Florida Cooperative Extension Service: USA, **1990**.

[38] Gallitelli, D. The ecology of Cucumber mosaic virus and sustainable agriculture. *Virus Res.,* **2000**, *71*(1-2), 9-21.
[http://dx.doi.org/10.1016/S0168-1702(00)00184-2] [PMID: 11137158]

[39] Rahman, M.S.; Ahmed, A.U.; Jahan, K.; Khatun, F. Management of cucumber mosaic virus (CMV) infecting cucumber in Bangladesh. *Bangladesh J. Agric. Res.,* **2022**, *45*(1), 65-76.
[http://dx.doi.org/10.3329/bjar.v45i1.59839]

[40] Draeger, K.R. Cucumber mosaic. University of Wisconsin-Madison. *Plant Pathol.,* **2021**, *D0036*, 1-4.

[41] Akanda, A.M.; Alam, N.; Khair, A.; Muqit, A. Altered metabolism of tomato leaves due to cucumber mosaic virus. *Bangladesh J. Sci. Res.,* **1998**, *16*(1), 1-6.

[42] Meiners, J.P.; Waterworth, H.E.; Smith, F.F.; Alconero, R.; Lawson, R.H. A seed-transmitted strain of cucumber mosaic virus isolated from bean. *J. Agric. Univ. P. R.,* **1977**, *61*(2), 137-147.

[43] Conti, M.; Caciagli, P.; Casetta, A. Infection sources and aphid vector in relation to the spread of Cucumber Mosaic Virus in pepper crops. *Phytopathol. Mediterr.,* **1979**, *18*, 123-128.

[44]   Mahjabeen, ; Akhtar, K.P.; Sarwar, N.; Saleem, M.Y.; Asghar, M.; Iqbal, Q.; Jamil, F.F. Effect of cucumber mosaic virus infection on morphology, yield and phenolic contents of tomato. *Arch. Phytopathol. Pflanzenschutz,* **2012**, *45*(7), 766-782.
[http://dx.doi.org/10.1080/03235408.2011.595965]

[45]   Khan, A.A.; Sharma, R.; Afreen, B.; Naqvi, Q.A.; Kumar, S.; Snehi, S.K.; Raj, S.K. Molecular identification of a new isolate of Cucumber mosaic virus subgroup II from basil (Ocimum sanctum) in India. *Phytoparasitica,* **2011**, *39*(2), 199-203.
[http://dx.doi.org/10.1007/s12600-011-0146-8]

[46]   Li, N.; Yu, C.; Yin, Y.; Gao, S.; Wang, F.; Jiao, C.; Yao, M. Pepper crop improvement against cucumber mosaic virus (CMV): A review. *Front. Plant Sci.,* **2020**, *11*, 598798.
[http://dx.doi.org/10.3389/fpls.2020.598798] [PMID: 33362830]

[47]   Bos, L.; Maat, D.Z. A strain of cucumber mosaic virus, seed-transmitted in beans. *Neth. J. Plant Pathol.,* **1974**, *80*(4), 113-123.
[http://dx.doi.org/10.1007/BF01981373]

[48]   Surasak, K. Persistence RNA viruses of common bean (Phaseolus vulgaris): Distribution and interaction with the host and acute plant viruses. Louisiana State University, Doctoral Dissertation. **2016**. Available from: https://digitalcommons.lsu.edu/gradschool_dissertations/838 (Date accessed: 07/06/2022).

[49]   Lipa, J.J. Compendium of bean diseases. Second edition. American Phytopathological Society, St. Paul, MN. *J Plant Protect Res,* **2005**, *45*(3), 180.

[50]   Hamed Derbalah, A.S.; Elsharkawy, M.M. A new strategy to control Cucumber mosaic virus using fabricated NiO-nanostructures. *J. Biotechnol.,* **2019**, *306*, 134-141.
[http://dx.doi.org/10.1016/j.jbiotec.2019.10.003] [PMID: 31593748]

[51]   Mandal, S.; Mandal, B.; Haq, Q.; Varma, A. Properties, diagnosis, and management of Cucumber Green Mottle Mosaic Virus. *Plant Viruses,* **2008**, *2*(1), 25-34.

[52]   Lee, G.H.; Ryu, C.M. Spraying of leaf-colonizing Bacillus amyloliquefaciens protects pepper from Cucumber mosaic virus. *Plant Dis.,* **2016**, *100*(10), 2099-2105.
[http://dx.doi.org/10.1094/PDIS-03-16-0314-RE] [PMID: 30682996]

[53]   Johary, T.; Dizadji, A.; Naderpour, M. Biological and molecular characteristics of Bean common mosaic virus isolates circulating in common bean in Iran. *J. Plant Pathol.,* **2016**, *98*(2), 301-310.

[54]   Worrall, E.A.; Wamonje, F.O.; Mukeshimana, G.; Harvey, J.J.W.; Carr, J.P.; Mitter, N. Bean common mosaic virus and bean common mosaic necrosis virus: relationship, biology, and prospects for control. *Adv. Virus Res.,* **2015**, *93*, 1-46.
[http://dx.doi.org/10.1016/bs.aivir.2015.04.002] [PMID: 26111585]

[55]   Feng, X.; Myers, J.R.; Karasev, A.V. Bean common mosaic virus isolate exhibits a novel pathogenicity profile in common bean, overcoming the bc-3 resistance allele coding for the mutated eIF4E translocation initiation factor. *Phytopathology,* **2015**, *105*(11), 1487-1495.
[http://dx.doi.org/10.1094/PHYTO-04-15-0108-R] [PMID: 26196181]

[56]   Mwaipopo, B.; Nchimbi-Msolla, S.; Njau, P.J.R.; Mark, D.; Mbanzibwa, D.R. Comprehensive Surveys of *Bean common mosaic virus* and *Bean common mosaic necrosis virus* and Molecular Evidence for Occurrence of Other *Phaseolus vulgaris* Viruses in Tanzania. *Plant Dis.,* **2018**, *102*(11), 2361-2370.
[http://dx.doi.org/10.1094/PDIS-01-18-0198-RE] [PMID: 30252625]

[57]   Prasad, H.P.; Shankar, U.A.C.; Kumar, B.H.; Shetty, S.H.; Prakash, H.S. Management of *Bean common mosaic virus* strain *blackeye cowpea mosaic* (BCMV-BlCM) in cowpea using plant extracts. *Arch. Phytopathol. Pflanzenschutz,* **2007**, *40*(2), 139-147.
[http://dx.doi.org/10.1080/03235400500356111]

[58]   Burke, D.W. Time planting in relation to disease-incidence and yields of bean in central Washington.

Plant Dis. Rep., **1964**, *48*, 789-793.

[59]   Wegulo, S.N.; Hein, G.L. Yellow dwarf of wheat, barley, and oats. NebGuide, University of Nebrasca-Lincoln Extension. *Institute of Agriculture and Natural Resources.,* **2013**, *G1823*, 1-4.

[60]   Walls, J.; Rajotte, E.; Rosa, C. The past, present, and the future of barley yellow dwarf management. *Agric,* **2019**, *9*(1), 23.
       [http://dx.doi.org/1-.3390/agriculture9010023]

[61]   Feng, X.; Guzmán, P.; Myers, J.R.; Karasev, A.V. Resistance to Bean Common Mosaic Necrosis Virus conferred by the bc-1 gene affects systemic spread of the virus in common bean. *Phytopathology,* **2017**, *107*(7), 893-900.
       [http://dx.doi.org/10.1094/PHYTO-01-17-0013-R] [PMID: 28475025]

[62]   Makkouk, K.M.; Kumari, S.G.; van Leur, J.A.G.; Jones, R.A.C. Control of plant virus diseases in cool-season grain legume crops. *Adv. Virus Res.,* **2014**, *90*, 207-253.
       [http://dx.doi.org/10.1016/B978-0-12-801246-8.00004-4] [PMID: 25410103]

CHAPTER 5

Economic Importance and Control of Vertebrate Pests in Legumes

Hafiz A. Badmus[1,2,*] and **Abideen A. Alarape**[3]

[1] *Department of Biodiversity, School of Molecular and Life Sciences, Faculty of Science and Agriculture, University of Limpopo, Limpopo Province, Republic of South Africa*

[2] *Department of Crop Protection and Environmental Biology, Faculty of Agriculture, University of Ibadan, Ibadan, Nigeria*

[3] *Department of Wildlife and Ecotourism Management, Faculty of Renewable Natural Resources, University of Ibadan, Ibadan, Nigeria*

Abstract: One of the constraints to crop production across the world is vertebrate pests. They have been implicated as the most destructive pests which inflict both pre-harvest and post-harvest damages on agricultural production. Legumes are one of the crops usually attacked by vertebrate pest species, though the degree of depredation varies from one crop type to another. Meanwhile, there has been a misconception among farmers, especially in some of the developing countries, that vertebrate pest species belonging to the order Rodentia are very difficult to control. This is evident in their crop cultivations whereby two rows are planted in addition to every eight rows of crop, for rodent pest species that may come and inflict damage on the cultivated crop. Some of the rodent pest species that cause economic damage to legumes on the field include *Arvicanthis niloticus*, *Xerus erythropus*, *Cricetomys gambianus*, *Rattus rattus*, *R. norvegicus*, and *Mus sp.*, while avian pest species include *Francolinus bicalcaratus*, and *Ploceus cucullatus*. There is a need to effectively manage these vertebrate pest species. Some of the rodent pest management approaches include the use of sanitation measures, exclusion of the vertebrate pest species, and modification of their habitat, and Trap Barrier System, while some of the avian pest management approaches include cage, nets or synthetic fibres, bird scarers, chemical repellents, sound-making devices, chemical poisoning, and trapping.

Keywords: Legumes, Rodent pests, Pre-harvest damage, Post-harvest damage, Avian pests, Management approach.

---

[*] **Corresponding author Hafiz A. Badmus:** Department of Biodiversity, School of Molecular and Life Sciences, Faculty of Science and Agriculture, University of Limpopo, Limpopo Province, Republic of South Africa; & Department of Crop Protection and Environmental Biology, Faculty of Agriculture, University of Ibadan, Ibadan, Nigeria; Tel: +234-803-826-7691; E-mail: badmus.hafiz@gmail.com

Phetole Mangena & Sifau A. Adejumo (Eds.)

INTRODUCTION

Vertebrate pests are pest animals characterized by the possession of a backbone. They are any vertebrate whether indigenous or exotic, wild or domestic, that has been implicated as the causes of economic, environmental, social, and health problems [1]. Species of vertebrates known to be pests could be found in any of the vertebrate classes which include amphibians, reptiles, birds, and mammals. However, class Mammalia has the highest pest species followed by class Aves [2]. Among the class Mammalia, rodent pest species have been identified as the most destructive categories of pests, globally [3]. Unfortunately, they are the most often overlooked pest species especially in developing countries and so are given little or no attention [4]. Even though a lot of people in developing countries share their insufficient households and diets with rodents and avian pest species, scientists and agriculturists are still not able to properly document quantitative losses by these vertebrate pest species. A twofold loss which comprised a percentage of their produce both at pre- and postharvest stages is suffered by the farmers' households [5]. Enormous amounts of produce damage and scarcities in some continents have been reported to be caused by vertebrate pest damage, particularly, by rodents [6]. Small mammals inflict a greater danger to crops of peasant farmers in Africa due to the injury and losses caused by them and their high costs of management compared to other countries worldwide [7].

Small mammals pose a significant constraint to crop production in agricultural ecosystems globally and managing them is still a major problem for researchers and agriculturalists. While few studies are relatively available to provide correct estimations of losses of crops due to rats in African countries, new research on farmer's familiarity, attitude, and practices in rat management showed that small mammals are regarded as the most persistent pest to manage [8, 9]. About 25 small mammal species have been recorded as pests in agriculture in African countries, causing different damage and losses in different crops [7]. According to the estimate, about one-fifth of the produce cultivated yearly worldwide are never consumed by individuals because of rodents-inflicted injury [10]. Aves can wreak injury to the vegetative and reproductive stages of all agricultural crops, starting from sowing, planting, and harvesting. Old-style methods usually rely on scaring birds by just rebounding the avian species to adjoining growing crops. However, it is an expensive management strategy [11].

RODENTS AND BIRDS AS VERTEBRATE PESTS OF LEGUMES

For small mammals (rodents and insectivores), damage to legumes is negligible [5]. Except for groundnuts, grain legumes are not the favorite foods of rats and mice. Most losses are not due to rodents but fungi and invertebrates of the class

Insecta. Skilled viewers come to an agreement that losses of legumes after harvest frequently surpass those of cereal crops. In addition, the avian damage to grain legumes is restricted to the field where avian pest species such as *Ploceus cucullatus* (weaverbirds) depredate the crop by removing the seeds from the pod [12]. Birds that do not live inside the farm or village structures like rodents, hardly ever depredate stored produce. Only the out-of-door conditions where cereals or legumes are unprotected during processing that aves can consume them, or they may have access to grain produce where they are kept under exposed storage conditions. Thus, the damage to stored produce because of avian activities is minimal compared to those caused by small mammals especially rodent species [1, 5].

IMPACT OF VERTEBRATE PESTS ON LEGUMES

Vertebrate pest species, especially those found in the classes Mammalia and Aves, inflict both pre-harvest and post-harvest damages to grain legumes [13]. Table **1** shows some of the legumes and the types of damage done by some vertebrate pests.

Table 1. Vertebrate pests and type of damage inflicted in some leguminous crops.

Crop	Type of Damage	Vertebrate Pest Indicted
Arachis hypogaea (groundnut)	Removal of newly sown and germinating seeds	Cane rat (*Thryonomys swinderianus*) Bush fowl (*Francolinus bicalcaratus*) Nile harsh-furred rat (*Arvicanthis niloticus*)
	Removal of pods	Red-legged ground squirrel (*Xerus erythropus*)
	Eating of roots and/or groundnut.	Mole rat (*Nesokia indica*)
	Removal of groundnut in the pod.	Lesser bandicoot rat (*Bandicota bengalensis*)
	Removal of groundnut but the plant is not usually damaged.	Indian gerbil (*Tatera indica*)
Vigna unguiculata (Cowpea)	Nibble on the cowpea grain in the store.	Mouse (*Mus minutoides*)
	Gnawing the stored cowpea.	Roof Rat (*Rattus rattus*) Norway Rat (*Rattus norvegicus*)
	Eating the seeds inside the pod.	Weaverbird (*Ploceus cucullatus*)
Pisum sativum (Garden pea)	Destruction of leaves, shoots, and mostly pods and seeds.	Rat (*Rattus sp*)

CONTROL OF VERTEBRATE PESTS OF LEGUMES

Rodent Pests

In any pest management system, the main objective is to reduce the damage caused by the pest species and not to destroy the animals because they have ecological services that they render in the ecosystem. Many vertebrate pest issues can be solved by applying precautionary approaches, such as hygiene, exclusion of pest, and habitation alteration [14].

Sanitation/Hygiene

Sanitation is the orderly arrangement and maintenance of a clean environment in and around warehouses and stores. Improvement of sanitation standards can prevent infestation and make the conditions unconducive for the pest species' survival [14].

Exclusion

Exclusion of rodent pests can be achieved by using mechanical proofing measures or physical barriers. The intention is to deny the vertebrate pest species access to food, water, shelter, and nesting places [14, 15]. Some of the barriers include electrical barriers, and frightening/ultrasonic devices among others.

Habitat Modification

Modification of the vertebrate pest habitat can also be achieved. This can be done by manipulating food and shelter to reduce rodent populations or the damage caused by them. There has been significant development of approaches to lessen populations and injury by manipulating vegetation. Many of these manipulations are not done just to reduce rodent habitation (which may be an accompanying benefit) but for other motives such as to reduce vegetative rivalry with crops or trees, to reduce soil pathogens, or to prepare the site for planting [15]. The use of decoy crops or additional feeding to lessen injury by small mammals or other vertebrates has also documented some success. Broken maize seed or soya beans have been spilled after drill-seeding on no-till cropland so that voles and other rodents will feed on those plants rather than feeding on recently developed crop plantlets or excavating and consuming the seed of the cultivated plant [16]. Fig. (1) below shows how and why animal populations grow and decline. This is regarded as population dynamics. Environmental factors (Habitat, predation and disease, competition) act through forces in the environment (reproduction, mortality, emigration, and immigration) to give rise to changes in the population at any point in time and space.

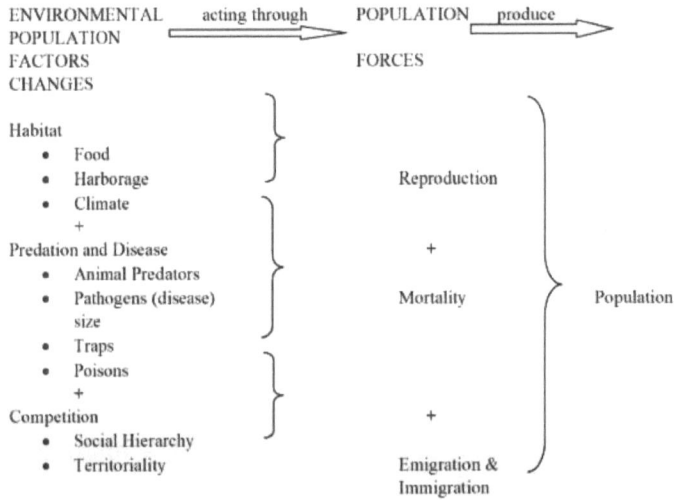

Fig. (1). Control mechanisms in animal populations [17].

Every environment has what we call the capacity which the environment can carry. The "carrying capacity" of the environment is defined as the number of animals that can be supported by the environment. Fig. (2) below shows the graph of an imaginary pattern of growth for a rodent population. It depicts an unchecked population that swings around a point of equilibrium (at "carrying capacity"). On the graph, there is an action level. The action level is the population size at which action must be engaged to avert public health, pest, or economic damage from happening. It is worthy of note that the number of animals the environment can support is above the "action level".

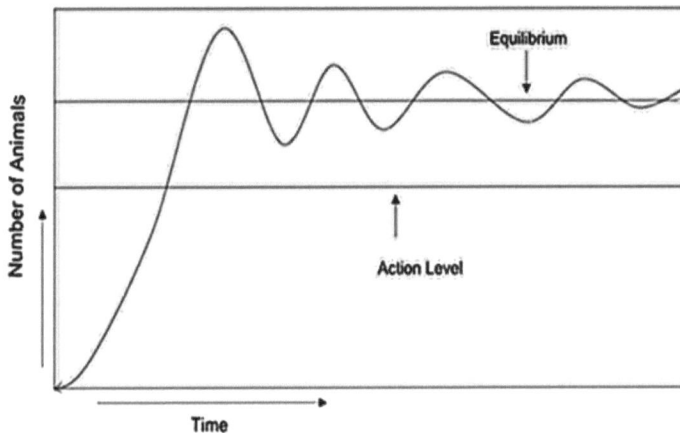

Fig. (2). Imaginary growth pattern for a rodent population [17].

Fig. (**3**) below shows the imaginary growth pattern for the impermanent lessening of the rodent population. The graph indicates a temporarily repressed population. Individual rodents are detached, but because the environment carrying capacity is unchanged, the population recoils to the level that can be supported by the environment.

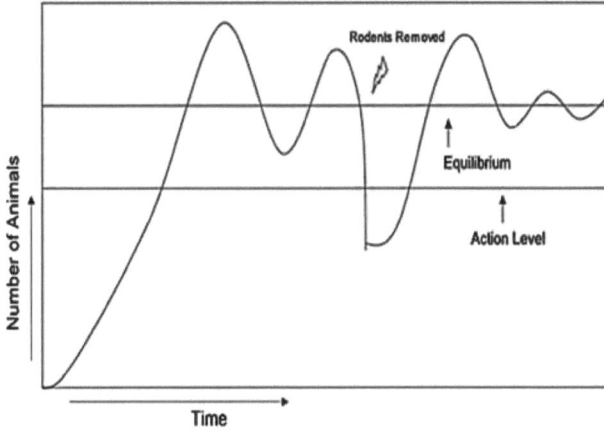

Fig. (3). Imaginary growth pattern for temporary reduction of rodent population [17].

Fig. (**4**) illustrates the imaginary growth pattern for a rodent population when its habitat is altered. This graph represents a lastingly suppressed population. This is the result of altering the habitat to support fewer rodents. This affects the whole population, including those that breed. There is now a new equilibrium that is below the level of worry, which is below the "tolerance limit".

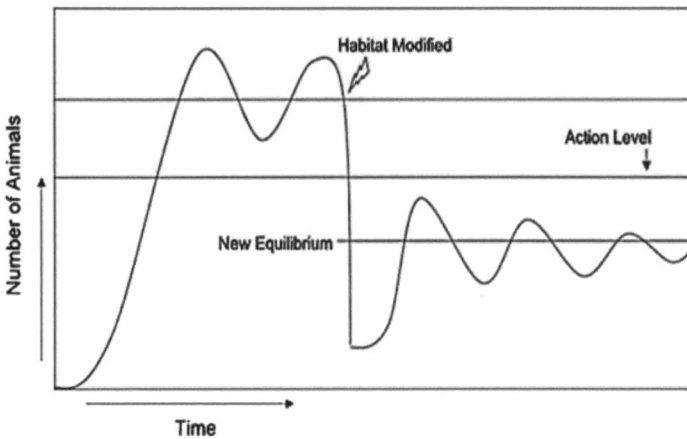

Fig. (4). Imaginary growth pattern for a rodent population when habitat is modified [17].

Trap Barrier System

Another method of rodent populace lessening is the rodent barrier method using a trap-barrier-system (TBS). TBS has been established that utilises some initially planted fields, cultivated to attract small mammals to them [3, 18]. The attracted fields are fenced by a rodent barrier, but they are regularly set apart apertures into wire snares to capture rodents. Occasionally, the wire snares are checked for any ensnared rodents. This method decreases rodent attack in the neighbouring crop fields that are cultivated [15].

Non-lethal Avian Control Methods

Cage Control

The screenhouse is taken to be an enclosure in the form of a cage meant for planting. Planting in the screenhouse averts attacks by birds. Screenhouses used for planting prevent the birds from gaining access to the planted crops. However, the construction of such screenhouses is expensive to set up and manage. Mostly valued crops, like those cultivated on investigation farmsteads for variety testing, hybrid growth or seed proliferation, can be fully protected from birds by being grown in cages [19]. They are costly and require a wood or a framework of metal supporting either wire or textile net with a small but sufficient net (20-25 mm) to exclude all seedeaters. This method of planting in cages or screenhouse is not meant for general farmers because of the cost-related issues. Meanwhile, farmers can go for a less operative but much less costly option which is the use of fibers and nets as a replacement for cages to prevent and keep the avian pests away from the crops [19, 20].

Loose Nets or Synthetic Strands

Nettings that are loose or artificial strands can be positioned straight onto a crop [19]. This must be cautiously done in order to stop aves from perching on or consuming the taller ears of the crop through the net. A modest building of posts and string like that used in the building of a cage but less intricate; will give improved support for the net and prevent birds, keeping them far from the farms. Synthetic fibres, such as acrylic fibre produced in the form of an entangled loosened net can also be used. Although, the fibre is not as costly as the nets, but it has been reported to be severely affected by heavy downpours and can be removed by strong winds. Elimination of the material from a crop needs care if it is to be utilised again. The fibre has advantages over the net in that it is not as expensive as netting. It can shield seedling plots that are not too large, especially whenever the birds are inflicting damage [21].

Bird Scarers

Another method of avian control that is non-lethal is the use of aves' scarers. The fright response of birds to scary shapes such as scarecrows has been traditionally applied. The extreme fright will make the aves to dodge the source of the trouble. To ensure the best guard to the crops, scare tactics of different varieties and irregularities should be adopted. However, due to the constant use of similar scaring tactics, this strategy no longer operates well except for a brief period. This is because the aves soon get familiar with the similar fright used over a long time [22]. Flapping, shouting, and running through the field are regarded as the modest form of aves scaring. In addition, it is required that the active bird scarer strengthens his screams with other sounds, such as the pounding of hollow tins as well as throwing missiles in the form of lumps of mud or gravel at any flock of aves trying to land in the field [19]. Scaring as a non-lethal method of avian control is more convenient to make by building a plinth on which the scarer stands to oversee the cultivated crop. The scarer of the aves sitting on his stage has a streak/line or connected to the system which, when pulled, agitates the rattles, and waves the flags. Dragging the line, screaming, and hurling gravel or sludge, when necessary, chase away the birds. This control method may lose its effectiveness if the aves are extremely hungry which makes them disregard the bird scarers [23].

Repellent Devices

The use of repellents is one of the scaring devices known to be easily applied and affordable. It is employed in certain conditions whereby black cotton or tape is threaded over the crop on the field. Communal aves release fright sounds whenever they collide with the thread and caution the congregation of birds. Its effectiveness against ploceids is low. The development of this method includes the use of flashing tapes which are overextended above the crop in parallel lines. The tapes (made of metals) are silver-coated on one side and shining on the other. So, whenever there is a bright light flashing on the tape, the tape twists and runs up and down. It also releases a noise in the breeze if stretched enough. The tape, however, only involves one of the bird's senses-sight. The method is stated here to accentuate the efficacy of the 'invisible' strands to which the aves cannot accustom [19, 24].

Sound-Making Devices

Another non-lethal method of avian control is the use of devices that make sound [18]. The efficacy of the methods that exploit sound is also limited because the avian species would adapt to the devices. Carbide guns or canons, produce bursts at consistent intermissions. Efficacy is, however, guaranteed as it increases if many sound devices are used. Some representations alternate and transform shooting direction repeatedly [19]. Efficacious devices that make noise have also been presented such as the 'A' alarm, an electronic amplifier that is battery-operated which makes complex irregular sounds that can be diverse in pulsation and pitch [25].

Chemical Repellent

Chemical repellents are odour-offensive chemicals used purposely to repel pest species. Chemical repellents that alter the taste of crops induce displeasures in granivorous aves feeding and this phenomenon is properly recognized in Africa mainly among the growers of millet and sorghum. Crop seeds contain variable amounts of tannins. Seeds having high tannin content that is high are usually evaded by avian species. Aves avoid such seeds, especially in situations where they can switch to an alternative seed crop if available. In places where there are severe aves problems, resistant varieties that are used for brewing and often for human consumption are frequently cultivated. Non-toxic but unpalatable chemicals keep birds away from the crop and injury. Methiocarb as the most effective material has reportedly decreased damage to grains by aves but also altered yield quantity and quality to an acceptable economic level. However, it has not been specifically approved for use because not too much work has been done on it. Chemical repellents can also be used as seed dressing. Thiram, a fungicide, is one of the chemicals that have been tested in Africa [4]. Although not promising, thiram use in dressing of seeds against fungi may be advantageous but it is slightly effective. Furthermore, the probability of the cost-effectiveness of chemical repellents in leguminous and non-leguminous crops remains uncertain [26].

LETHAL CONTROL

Assassination of birds in the crop is another approach occasionally used by farmers. Birds are killed, either by poisoning or by trapping.

Killing of Birds Using Chemical Poisoning

According to Shefte *et al.* [4], most of the damage done to crops on the field by avian pests is around the edges of the cultivated fields, especially in places near the cover (trees or shrubs). A swath about 5 to 10 m into the crop is sprayed with a pesticide identified to be toxic to aves so that any avian pest species that comes around to eat the crop that has been sprayed is killed. However, it is important that the pesticide must be persistent to be effective so that spraying would not be repeatedly carried out. It is strongly advised that pesticides with persistence features should be discouraged in grain crops, especially near harvest periods if the grain is meant for animal or human consumption [27].

Killing of Birds by Trapping

Avian species can be controlled by trapping. Nevertheless, this method is not very efficacious in large field conditions [19]. The use of crow trap has been experimented with in Sudan whereby the trap is baited with grain and live decoy aves. In addition, the quelea colony has been efficaciously controlled in Chad using water as bait. Wire cages of $1m^3$ in size were set, with a drooping sheet of plastic material as the top side of the cage. A slit was then created in the sheet in such a way that the landing of the birds on the surface would make them slide inescapably through the slit into the trap and would not be able to escape. Although, the trap is not yet in use in many crop fields, this approach is cheap to construct and recyclable [19, 28].

CONCLUSION

Legumes are one of the crops usually attacked by vertebrate pest species. The vertebrate pest species that have been implicated in the legume attack are either rodent or avian species. The rodent damage to grain legumes is minimal. Some of the rodent pest species that cause damage to legumes include *Arvicanthis niloticus*, *Xerus erythropus*, *Cricetomys gambianus*, *Rattus rattus*, *R. norvegicus*, and *Mus sp.* The damage done by birds to grain legumes is limited to the field where avian pest species such as weaverbirds depredate the crop by removing the seeds from the pod. Some of the avian pest species that inflict damage on legumes include *Francolinus bicalcaratus*, and *Ploceus cucullatus.* However, rodent pest management strategies include the use of sanitation measures, exclusion of vertebrate pest species, and modification of their habitat, as well as the Trap Barrier System. Avian pests can be managed using cages, nets or synthetic fibres, bird scarers, chemical repellents, sound-making devices, chemical poisoning, and trapping.

CONSENT FOR PUBLICATION

Not applicable.

CONFLICT OF INTEREST

The authors declare no conflict of interest, financial or otherwise.

ACKNOWLEDGEMENT

Declared none.

REFERENCES

[1] Buckle, A.P.; Smith, R.H. *Rodent Pests and their Control,* 2nd; CPI Group (UK) Ltd, Croydon, CR0 4YY, **2015**.
[http://dx.doi.org/10.1079/9781845938178.0000]

[2] Happold, D.C.D. *Mammals of Africa. Volume III: Rodents, Hares and Rabbits*; Bloomsbury Publishing: London, **2013**.

[3] Aplin, K.P.; Brown, P.R.; Jacob, J.; Krebs, C.J.; Singleton, G.R. Field methods for rodent studies in Asia and the Indo-Pacific. *ACIAR Monograph,* **2003**, (100), 223.

[4] Shefte, N.; Bruggers, R.L.; Schafer, E.W., Jr Repellency and toxicity of 3 control chemicals to for species of Africa grain-eating birds. *J. Wildl. Manage.,* **1982**, *46*(2), 453-457.
[http://dx.doi.org/10.2307/3808656]

[5] Brooks, J.E.; Fiedler, L.A. *Vertebrate Pests: Damage on Stored foods*; USDA/APHIS/WS/National Wildlife Research Center: Fort Collins, CO 80524 USA, **1999**, p. 27.

[6] Fayenuwo, J.O.; Olakojo, S.A.; Akande, M.; Amusa, N.A.; Olujimi, O.A. Comparative evaluation of vertebrate pest damage on some newly developed quality protein maize (QPM) varieties in south western Nigeria. *Afr. J. Agric. Res.,* **2007**, *2*(11), 592-595.

[7] Makundi, R.H.; Oguge, N.O.; Mwanjabe, P.S. Rodent pest management in East Africa : An ecological approach. In: *Ecologically based Rodent Management*; Singleton, G.; Hinds, L.; Leirs, H.; Zhang, Z., Eds.; Australian Centre for International Agricultural Research: Canberra, **1999**; pp. 460-476.

[8] Makundi, R.H.; Bekele, A.; Leirs, H.; Massawe, A.W.; Rwamugira, W.; Mulungu, L.S. Farmer's perceptions of rodents as crop pests: Knowledge, attitudes, and practices in rodent pest management in Tanzania and Ethiopia. *Belg. J. Zool.,* **2005**, *135* Suppl., 153-157.

[9] Yonas, M.; Welegerima, K.; Deckers, S.; Raes, D.; Makundi, R.; Leirs, H. Farmers' perspectives of rodent damage and management from the highlands of Tigray, Northern Ethiopian. *Crop Prot.,* **2010**, *29*(6), 532-539.
[http://dx.doi.org/10.1016/j.cropro.2009.12.006]

[10] Sarwar, M. Pattern of damage by rodent *(Rodentia: Muridae)* pests in wheat in conjunction with their comparative densities throughout growth phase of crop. *Int. J. Scient. Res. Environ. Sci.,* **2015**, *3*(4), 159-166.
[http://dx.doi.org/10.12983/ijsres-2015-p0159-0166]

[11] Bruggers, R.L. The situation of grain eating birds in Somalia. *In: Proceedings of the 9th Vertebrate Pest Conference.,* The situation of grain eating birds in Somalia. Frenno, Carlifornia, 1980, pp. 5–16.

[12] Brooks, J.E.; Ahmad, E.; Hussain, I. Damage by vertebrate pests to groundnut in Pakistan. *Proceedings of the 13th Vertebrate Pest Conference,* **1988**, , pp. 129-133.

[13] Brown, P.R.; McWilliam, A.; Khamphoukeo, K. Post-harvest damage to stored grain by rodents in village environments in Laos. *Int. Biodeterior. Biodegradation,* **2013**, *82*, 104-109.
[http://dx.doi.org/10.1016/j.ibiod.2012.12.018]

[14] Ofuya, T.I.; Lale, N.E.S. *Pests of stored cereals and pulses in Nigeria: Biology, ecology and control*; Dave Collins Publications: Nigeria, **2001**, pp. 133-146.

[15] Witmer, G.W. *The ecology of vertebrate pests and integrated pest management (IPM)*; USDA National Wildlife Research Center - Staff Publications, **2007**, p. 730.
[http://dx.doi.org/10.1017/CBO9780511752353.013]

[16] Witmer, G.W.; VerCauteren, K.C. Understanding vole problems in direct seeding : Strategies for management. *Northwest Direct Seed Conference,* Spokane, **2001**, pp. 104-10.

[17] Bio-Integral Resource Centre. *Commensal Rodents – Biology, Population Dynamics and IPM*; Curriculum for Pest Management Professionals. Global Management Options, **2005**, p. 35.

[18]   Singleton, G.R.; Sudarmaji, S.; Suriapermana, S. An experimental field study to evaluate a trap-barrier system and fumigation for controlling the rice rat in rice crops in West Java. *Crop Prot.,* **1998**, *17*, 55-64.
[http://dx.doi.org/10.1016/S0261-2194(98)80013-6]

[19]   Maurice, M.E.; Fuashi, N.A.; Mengwi, N.H.; Ebong, E.L.; Awa, P.D.; Daizy, N.F. The control methods used by the local farmers to reduce Weaver-Bird raids in Tiko Farming Area, Southwest Region, Cameroon. *Madridge. J. Agricul. Environ. Sci.,* **2019**, *1*(1), 31-39.
[http://dx.doi.org/10.18689/mjaes-1000106]

[20]   Gadd, P., Jr Use of the Modify ed Australian Crow (MAC) trap for the control of depredating birds in Sonoma County. *Proceedings of the 17th vertebrate pest conference,* University of California, Davis, 1996, pp. 103-107.

[21]   Lowe, K.W. *The Australian Bird Bander's Manual*; Australian National Parks and Wildlife Service: Canberra, **1989**.

[22]   Bishop, J.; McKay, H.; Parrott, D.; Allan, D. *Review of international research literature regarding the effectiveness of auditory bird scaring techniques and potential alternatives*; Department for Environment, Food and Rural Affairs: New York, **2003**.

[23]   GTZ. *The ecology and control of the Red-billed weaver Bird Quelea quelea L. in Northeast Nigeria. Special publication No. 199. Eschborn*; Deutsche Gesselschaft fur Technische Zusammenarbeit (GTZ): West Germany, **1987**.

[24]   Meinzingen, W.F. New Application Techniques. Internal report of FAO/ UNDP Regional Quelea Project RAF/81/023. Monitoring. *In: Proceedings of Vegetation 2000, 2 years of operation to prepare the future,* Lake Maggiore – Italy**1984**.

[25]   Marsh, R.E.; Erickson, W.A.; Salmon, T.P. *Bird Hazing and Frightening Methods and Techniques*; Other Publications in Wildlife Management, **1991**.

[26]   Mason, J.R.; Clark, L. Avian repellents: options, modes of action, and economic considerations. In: *Repellents in Wildlife Management*; National Wildlife Research Centre: Fort Collins, Colorado, **1997**; pp. 371-391.

[27]   Pope, G.G.; Ward, P. The effects of small application of an organophosphorus poison, fenthion, on the weaver bird, *Quelea quelea. Pest Manag. Sci.,* **1972**, *3*(2), 179-205.

[28]   Van Vuren, D.; Smallwood, K.S. Ecological management of vertebrate pests in agricultural systems. *Biol. Agric. Hortic.,* **1996**, *13*(1), 39-62.
[http://dx.doi.org/10.1080/01448765.1996.9754765]

CHAPTER 6

Effect of Spider Diversity and Abundance in Legume Agroecosystems

Mokgadi Asnath Modiba[1,*], **Sinorita Chauke**[1] and **Yolette Belinda Rapelang Nyathi**[1]

[1] Department of Biodiversity, School of Molecular and Life Sciences, Faculty of Science and Agriculture, University of Limpopo, Limpopo Province, Republic of South Africa

Abstract: An agroecosystem refers to a complex system comprising a couple of different interacting factors, involving species, ecological, and management processes. This system contains lesser species diversity of both plants and animals than a natural ecosystem. The variation in species of plants and insects is critically important to serve as a complex food chain and web whose interactions function to stabilise this ecological unit. However, among the groups of herbivores and predators found in agroecosystems, spiders play a key role in most crop fields by preying on a variety of pests. Besides this, the current pace of research on this subject shows that the role of spiders in regulating pest species and serving as potential biological control agents has been largely ignored. So far, information on agricultural spider communities, diversity and their role as biological pesticides remain scant in various parts of the world with the exception of countries such as the United States of America, Australia, and some parts of the Middle East Asia. Thus, this chapter outlines the most relevant information on the diversity, abundance and effect of arthropodous spiders on agroecosystems, particularly those that are involved in the cultivation of legume crop species. The paper also discusses current relevant threats to spiders, conservation measures, the threat of species extinction, and the role that these arthropods play in agriculture, especially by reducing the growth and productivity of species such as soybean (*Glycine max* L.) and cowpea (*Vigna unguiculata*).

Keywords: Agroecosystems, Arthropods, Legumes, Soybeans, Spiders.

INTRODUCTION

An agroecosystem refers to a complex system comprising a couple of different interacting factors, involving species, ecological, and management processes [1]. This system contains lesser species diversity of both plants and animals than a natural ecosystem. Typically, one to four major crop species and six to ten major

* **Corresponding author Mokgadi Asnath Modiba:** Department of Biodiversity, School of Molecular and Life Sciences, Faculty of Science and Agriculture, University of Limpopo, Limpopo Province, Republic of South Africa; Tel: +2715-268-4045; E-mail: mokgadi.modiba@ul.ac.za

Phetole Mangena & Sifau A. Adejumo (Eds.)

pest species can be found in this ecosystem [1, 2]. In most cases, the status and conditions of agroecosystems are largely influenced by anthropogenic activities. Some of these activities include ploughing, inter-cultivation, and application of pesticides, leading to the alteration of the diversity of species, especially of pests. However, influences by man can be more detrimental, causing the agroecosystem to be more susceptible to pest damage and catastrophic outbreaks that are concomitantly attributed to the lack of species diversity. The variations in species of plants and insects are critically important to serve as a complex food chain and web whose interactions function to stabilise this ecological unit [2].

The conversion of ecological units to agriculture also leads to the invasion by unplanned diversity of weed plants, herbivores, predators, microbial pathogens, and other organisms that persist in the system. Among the group of herbivores and predators found in agroecosystems, arthropods spiders play a key role in most crop fields by preying on a variety of pests. However, the current pace of research on this subject shows that the role of spiders in regulating pest species and serving as potential biological control agents has been largely ignored. So far, information on agricultural spider communities, diversity and their role as biological pesticides remain scant in various parts of the world except for countries such as the United States of America, Australia, and some parts of Middle East Asia [3]. The functions of spiders as pest predators for herbivores and granivores remain promisingly beneficial for agriculture and offer an alternative pest management strategy for both small- and large-scale farmers.

Isbister [4] emphasised on the myths and reputation of spiders as also being the "predators of man" or being dangerous to people and animals as one of the main reasons why credit is still due to them for use as important natural pests control agents. But among over 30,000 known spider species, only about twenty-three of species are considered poisonous. As a result, this chapter outlines the most relevant information on the diversity, abundance, and effect of arthropodous spiders on agroecosystems, particularly those that involve the cultivation of legume crop species. The paper also discusses current relevant threats to spiders, conservation measures, threat of species extinction and the role that these arthropods play in agriculture, especially by reducing the growth and productivity of species such as soybean (*Glycine max* L.) and cowpea (*Vigna unguiculata*).

BIOLOGY OF SPIDERS

Structural Morphology and Life Cycle

Arthropods are invertebrates that form a significant part of the animal kingdom. They are easily identified and distinguished by the distinct morphological traits that they possess. Their features include an exoskeleton, paired jointed appendages and a segmented body. They possess abilities to survive in aerial, aquatic and terrestrial environments. For instance, the class Arachnida consists of eleven (11) orders of joint-legged invertebrates that also include spiders [5]. However, compared to other invertebrate animals, spiders form part of the phylum Arthropoda, subphylum Chelicerata [6]. These species are further classified into the order Araneae which consists of 112 families, 4072 genera with approximately 47000 species. The suborder Araneae are classified into two suborders known as the Mesothelae which consists of one family with 87 species which are characterized by traces of segmentation on their abdomen and the Opisthothelae which have no traces of segmentation on their abdomen [7, 8]. The biology of spiders indicates that they all undergo the same general stages of development. In general, many species go through the egg, spiderling, and adult stages as indicated in Fig. (1). But, having stated that slight differences in their developmental stages may obviously exist based on species variation within the taxa. During the egg or embryonic stage, the female spider builds an egg sac using silk and deposits her eggs inside it and fertilize them as they emerge. One egg sac may host up to a hundred eggs depending on the species. The eggs usually take a week to hatch but some spiders, especially those that are found in temperate regions, may employ specific strategies and other unique characteristics such as overwintering of the egg sacs, and then emerge in spring. Some spider species protect their egg sacs while others abandon the egg sacs in secure places. The spiderling stage commences as soon as they hatch from their eggs [9]. At this stage, the spiders are much smaller in size and immediately disperse through a process known as ballooning or walking. Most species become mature after shedding at least ten times. Males are usually fully mature by the time they leave the sac, but female spiders take more time to mature since they are usually larger than male spiders. In the adult stage, spiders become fully developed for mating, and after the events of mating between the two, female spiders will then live and survive for a longer period than male spiders which usually die after this process. The life span of spiders can be up to two years, but variability also exists between different species [3, 9].

Spiders consist of chelicera, which is a pair of appendages in the front of their mouth that allows for tearing apart of their prey instead of chewing. Furthermore, these chelicerae also contain two fangs at their tip which are connected to spiders' poisoning glands [10, 11]. The body of spiders is also divided into prosoma and opisthosoma as previously indicated. The prosoma consists of the eyes and locomotory appendages while the opisthosoma consists of the abdomen. Spinnerets and silk glands secrete silk which is a significant distinguishing character of spiders as earlier described by Haupt [10]. According to Haupt [10], Araneae also contains within its species some of the most venomous web-spiders whereby the venomous capabilities of some of species remain unknown. Spiders have varying morphological traits ranging from large to small-sized bodies, eight to two-eyed, very long to very short-legged, hairy to bald, using both or neither the book lung nor trachea for breathing. Some of these morphological features are represented in Fig. (**2**) [12].

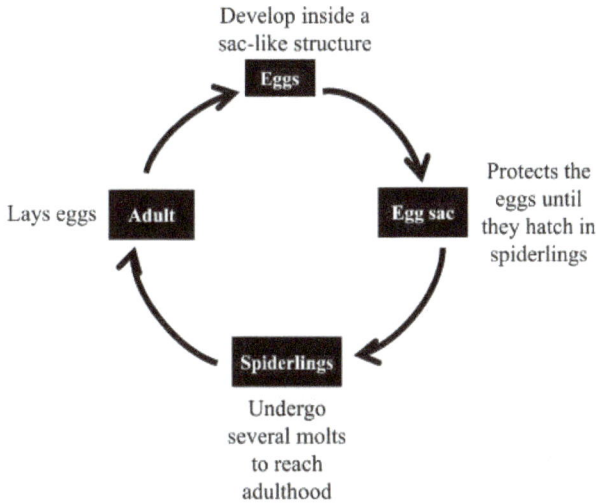

Fig. (1). General life cycle of spiders showing the three developmental stages: eggs, spiderling, and adult.

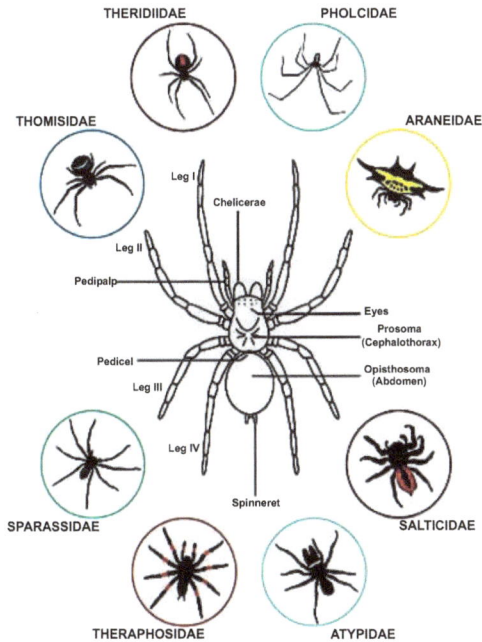

Fig. (2). Structural representatives of different spider families found under different domestic and natural habitats [7].

Spiders are also referred to as model predators due to their ability to devise varying strategies to trap prey. Some of their strategies include trapping prey in sticky webs, impersonating the prey, lassoing with sticky balls, sensing vibrations, and utilising their very sharp vision. However, spiders are also prey in the sight of other predators. Hence, they possess several mechanisms of defence such as the ability to camouflage, flicking hairs to their predators, trapping predators in their webs, and using venom, which is so far, their strongest form of defence [13]. Almost all spiders have venom glands that produce the poisonous substance that they inject into their prey through chelicerae. But most of them are not harmful to humans, apart from a few spiders containing venom that can be harmful and potentially cause death if the treatment is delayed [14]. The brown recluse spider, hobo spider, black widow spider, Armadeiras, funnel-web spider and Tarantula cause the most important syndromes such as loxoscelism, latrodectism and funnel web spider syndrome in humans [15]. Spiders feed on ants, bees, beetles, butterflies, earwigs, flies, harvestmen, wasps, woodlice, and other spiders. In rare cases, they are known to feed on caterpillars, pupae, worms, and small fish. Some large spiders feed on mice and snakes including *Thalassiiis spencer* which feeds on frogs and some species of *Mygalomorphae* which feed on birds [16].

IDENTIFICATION AND DISTRIBUTION

The identification and abundance of spiders involve the use of webs as one of the characteristics required to identify spiders. However, some spiders form webs while others do not build webs. In this case, spiders that construct webs are distinguished by the structure of their webs. Among these, cobwebs or tangle webs (*Theridiidae*), orbwebs (*Araneidae*) and sheet webs (*Linyphiidae*) have been identified [17]. Spider web designs also simplify the identification of other factors such as the prey, habitat, and predators. Spiders can make adjustments to their web structure in response to environmental changes [18]. Furthermore, the movement of the chelicera of spiders is used to differentiate between the three main groups of spiders (Araneomorphae, Mesothelae and Mygalomorphae). Mygalomorphs have fangs that move up and down while Araneomorphs have fangs that open laterally and move sideways [11]. Foraging strategies also serve as traits used to differentiate between spiders (Fig. **2**).

They are classified into stalkers (Salticidae and Oxyopidae), ambushers (Thomisidae and Pisauridae), foliage runners (Anyphaenidae and Clubionidae), ground runners (Lycosidae and Gnaphosidae), funnel web-builders (Agelenidae and Amaurobiidae), wandering sheet/tangle weavers (Linyphiidae), orb weavers (Araneidae, Tetragnathidae, and Uloboridae), and 3D web builders (Theridiidae and Pholcidae) based on hunting strategies [19, 20]. Uetz *et al.* [21] proposed a guild structure of spiders according to foraging strategies and the types of webs constructed by spiders as demonstrated in Fig. (**3**). However, the identification of spiders using morphological terms is time-consuming due to insufficient identification keys, polymorphism, and sexual dimorphism. Moreover, DNA barcoding has proved to be a convenient alternative method for precise identification of species [22]. Different species of spiders have varying requirements in terms of humidity and temperature preferences.

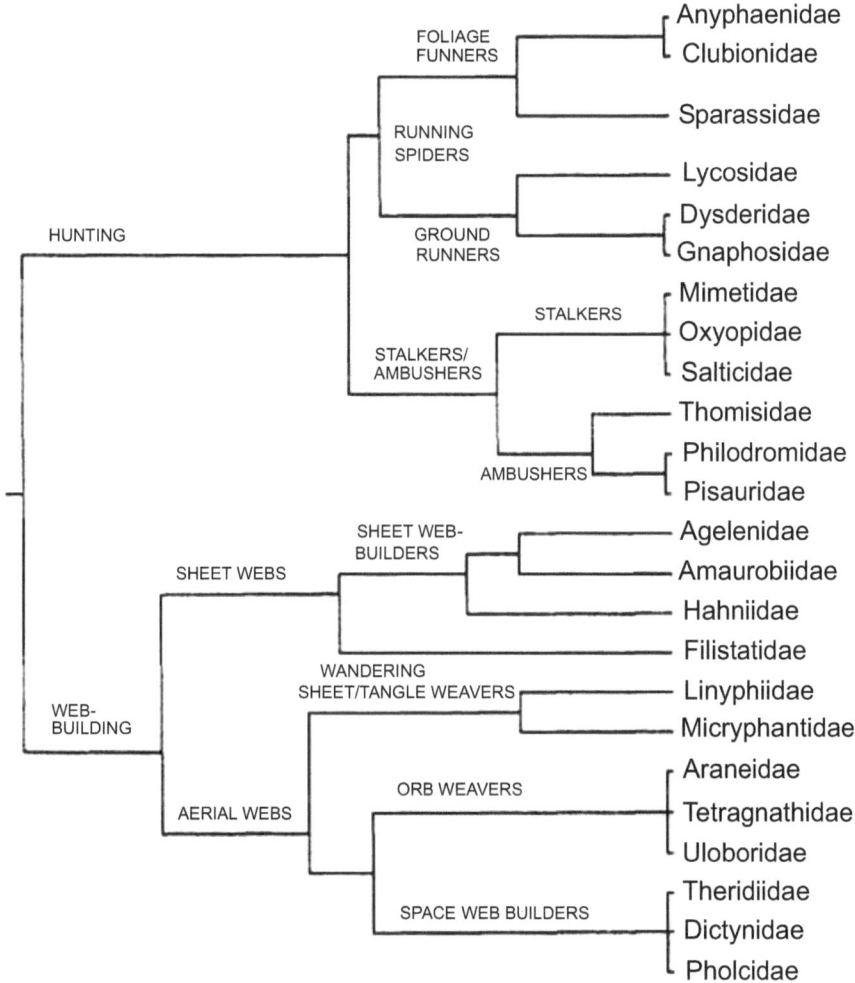

Fig. (3). Classification of spiders according to foraging strategies and types of webs [21].

Most of them are limited to environmental conditions that provide a microclimate within their physiological tolerance ranges [23]. However, some species are generalists and do not have specific habitat requirements [24]. However abiotic factors and seasonality have been shown to influence the distribution of species. A spider's choice of habitation also depends on various factors such as the abundance of prey, substrate availability, the presence of conspecifics, and other factors that may induce changes in a site or cause damage to their webs. As previously indicated, spiders disperse as spiderlings through the process called ballooning where they release single strands of silk which are caught by the wind and launched into the air. In this case, spiders have no control over where they land [25].

ABUNDANCE AND VALUE OF SPIDERS IN AGRICULTURE

Spiders are highly diverse, insecticide resistant, less vulnerable to changing climatic conditions, easily adapted to new climates and the most dominant insectivores in many agroecosystems [26]. Plant vegetation provides an environment for spiders to construct webs, hunt, and attract prey. Different species of spiders inhabit different plant parts, vegetation types, or certain plant species for varying purposes. Some spiders use the leaves on plants or dried leaves on the ground for hunting, shelter, and reproduction while others may use certain types of stems or branches for the same purposes and camouflaging. The characteristics that the plants possess also have an influence on determining the type of spiders that are likely to inhabit the plant. Certain species of spiders associate themselves with specific plant species or plants that have common morphological traits such as glandular trichomes and rosettes. Some spiders provide nutritional benefits to plants while also deterring herbivores.

They serve as biological control agents of pests in agroecosystems. They feed on insect herbivores thereby minimising herbivory on plants. The most common inhabitants of vegetation are the families of spiders that make up the guilds of stalkers, ambushers, and foliage runners [20]. According to Oberg [16], the different species of spiders are also known to be the most efficient in suppressing insect pests in crop fields. Although most spiders are regarded as obligatory carnivores, some spiders can also feed on vegetation and are regarded as omnivores. Over 60 species belonging to 10 families of spiders feed on plant products such as nectar and pollen. The families include the web-weaving (Araneidae, Linyphiidae, and Theridiidae) and non-weaving (Anyphaenidae, Clubionidae, Eutichuridae, Oxyopidae, Salticidae, Thomisidae, and Trachelidae) as earlier described by Oberg [16] and recently by Vasconcellos-Neto *et al.* [13].

IMPACT OF SPIDERS ON LEGUME CROPS

The microclimate provided by soybean crops is very favourable for spider multiplication, and in turn, spiders serve as biopesticides for insect pests affecting legumes [26]. The presence of spiders reduces the foraging activities of pest insects on leguminous crop leaves [13, 20]. In soybean fields, for example, this crop hosts more diverse communities of spiders than any other agricultural crop fields [17]. Pearce *et al.* [24] observed species belonging to 27 spider families in soybean fields. Some of these families are summarised in Table **1**, together with the morphological attributes that are commonly used for spider species identification. The second table (Table **2**) below outlines some of the most common habitats in which spider species within these families are also found, including in agricultural fields. When spider communities are found inhabiting agroecosystems or agricultural lands where crops such as legumes are planted, they mostly help to carry out pest control by feeding on the insect pests.

Table 1. Morphological characteristics of some of the most common spider families found in soybean fields [21].

Family	Morphological Attributes
Agelenidae	½ to ¾ inch long body, scissor-like jaw, 8 small eyes arranged in a D form, long abdomen with a pair of long spinnerets, yellowish-tan and brown coloured, long, hairy and spiny legs
Araneidae	About one-third to one-inch long, scissor-like jaws, 8 small eyes spread apart, legs banded and spiny and round to heart-shaped abdomen.
Araneidae, Genus- Argiope	Their bodies are up to 1 inch long, have scissor-like jaws, 8 small eyes spread apart, and oval to shield shaped abdomen with yellow, black, and silver-white markings.
Clubionidae Corinnidae	1/10 to ½ inch long body, eight small eyes divided into rows, long oval-shaped abdomen, body, and legs covered with hairs, and lack, brown, tan or yellow coloured with light or dark markings.
Ctenidae	¼ to 1 inch long body, scissor-like jaws, small four anterior eyes and four larger posterior eyes, black, brown, tan, grey or yellow with light or dark markings, long hairy legs and some have fangs and anterior carapace.
Dysderidae	½ inch long body, 6 small eyes arranged in three pairs, abdomen underside has four respiratory slits, arranged in pairs on either side and long scissor-like jaws.
Filistatidae	1/3 to ¾ inch long body, have scissor-like jaws, 8 eyes with 4 anterior eyes larger than 4 posterior eyes, long and velvety legs with long pedipalps, shiny carapace with sparse hairs, oval-shaped and velvety abdomen.
Gnaphosidae	More or less than ½ inch long body size, scissor-like jaws, black or brown coloured with light markings, long abdomen with a pair of long spinnerets, eight closely assembled small eyes, velvety or glossy legs and carapace.
Lycosidae	¼ to 1 1/3 inches long body, 4 small anterior eyes, large posterior median eyes, and smaller posterior lateral eyes, scissor-like jaws, black, brown, yellow, grey, tan or yellow coloured with lighter or darker markings and long hairy legs.
Miturgidae	1/10 to ½ inch long body, eight small eyes divided into rows, long oval shaped abdomen, body and legs covered with hairs, and lack, brown, tan or yellow coloured with light or dark markings.

(Table 1) cont.....

Family	Morphological Attributes
Nephilidae	¾ to 1 inch long body, jaws are scissor-like, 8 small eyes, legs have a fuzzy-kneed appearance, abdomen is golden yellow, symmetrically dotted with small white spots and the legs are yellow and black in colour.
Pisauridae	½ to 1 inch long body, 4 small anterior and four larger posterior eyes, scissor like jaw, long hairy legs, long abdomen and tan or beige and brown coloured.
Sicariidae	1/3 inch long body, scissor-like jaws, long and slender legs, six small eyes grouped into three pairs, light brown oval shaped abdomen and dark brown violin pattern on light brown carapace.
Sparassidae	¾ to 1 inch long body, scissor-like jaws, long oval-shaped abdomen, brown and beige coloured, long legs with hairs and spines, small anterior and posterior median eyes, large anterior and posterior lateral eyes.
Tetragnathidae	1/3 to ½ inch long body, scissor-like jaws, 8 small eyes, long, thin, spiny, and translucent legs, long oval abdomens, black, silvery-white, yellow, orange, brown and green coloured.
Theridiidae	½ inch long body, scissor-like jaws, legs spindly, red, orange, or yellow coloured hourglass marking is apparent on the ventral surface of the abdomen.
Theraphosidae	1.5 to 2 inches long, jaws move up and down, body and legs usually hairy and black, brown and beige coloured.
Thomisidae	1/10 to ½ inch long body, scissor-like jaws, eight small eyes, legs with hairs and spines, front pair longer than hind pair, brightly coloured or combinations of black, gray, white, brown, rust, beige, and yellow.

Table 2. Description of habitats used by some of the most common spider families and species found in agricultural fields [21, 23].

Family	Habitat
Agelenidae	May be found close to building foundations or shrubs, ivy, tall grass, decks, and in window wells and recessed vents. Indoors, webs are found in the corners of garages, basements, and crawlspaces.
Araneidae	Form orb-shaped webs across porch balustrades, posts, doorways, windows and between branches.
Araneidae, Genus- Argiope	They form orb-shaped webs between branches of tall woody-stemmed weeds and manmade structures such as buildings.
Clubionidae Corinnidae	Beneath stones, folded leaves, corners, and dark environments indoors and folds of fabric.
Ctenidae	Found in vegetation and structures or on the ground.
Dysderidae	Live under stones, logs, and barks.
Filistatidae	Exterior walls of rustic outbuildings, barns, and abandoned and unkempt homes, as well as undisturbed attics and basements.
Apodidae	Beneath stones and loose barks, and leaf litter.
Lycosidae	Beneath stones and debris.
Miturgidae	Beneath stones, folded leaves, corners and dark environments indoors and folds of fabric.

(Table 2) cont.....

Family	Habitat
Nephilidae	Woody and low vegetation habitats, marshy areas, and bodies of water, trees, and shrubs near buildings.
Pisauridae	Vegetations exposed to sunlight, scarred or hollowed tree trunks, aquatic vegetation and cattail stems.
Sicariidae	Attics, crawl spaces, basements, wall voids, upholstered furniture, clothing, among items stored in cardboard boxes indoors and under stones in warm climates outdoors.
Sparassidae	Rest in indoor and outdoor crevices.
Tetragnathidae	Tall vegetations, shrubs, exterior features of buildings and warm climates.
Theridiidae	Corners of porticos, eaves, garages, windows and all indoor areas. Outbuildings, sheds, pump and meter enclosures commonly, building foundation perimeters., under rocks, in hollow trees stumps and among tree bark.
Theraphosidae	They live on the ground and burrows.
Thomisidae	flowers, leaves, bark, the ground, structural surfaces and crevices.

This kind of spider activity becomes more advantageous to farmers since crop productivity of the cultivated crops may significantly increase in the absence of insect pests and aphids that would normally feed on the crops. The biotic stress effects may be severe, especially if there are no carnivorous spiders and/ or any form of pest control measure imposed on them. Among the common groups, as illustrated in Table **1**, spider families that feed on the insects and pests include Thomisidae (Crab spiders), Salticidae (Jumping spiders) and Lycosidae (Wolf spiders) to mention just a few. These spiders remain active hunters of insects in vegetation gardens and crop fields [27]. As such, it is, therefore, clear that spiders are highly needed in the development of sustainable and low-input agricultural ecosystems that are efficient, particularly based on their important habits of feeding on the insects that affect crops.

Furthermore, spiders could make such positive impacts on crop productivity because they do not only feed on whole insects but also on the eggs and larvae of the pests. In addition to the positive impact that spiders have on legumes with respect to the pest population, spiders can also influence the composition of other natural enemies [27, 28]. The competitive interactions between spiders and their enemies such as fire ants can either have positive or negative effects on pest control which in turn affects legume crop productivity. In cases where fire ants might overpower the spiders, they may not be as effective as spiders with regard to the removal of harmful pests to crops, a competition that is also dependent on the time and season [23]. According to a report by Sunderland *et al.* [29], spiders can be very active in fields where legume crops such as alfalfa, soybeans and chickpea are cultivated. These crops are usually infested by many insect pests; thus, spiders are very effective in their pest control activities as they serve to patrol the croplands in search of insects for food. However, this can be both cost-effective and environmentally friendly as compared to the use of pesticides or insecticides [30].

OTHER PROBLEMS ASSOCIATED WITH SPIDER INFESTATIONS

Apart from spider mites (*Tetranychus urticae* Koch), typical spiders have not been directly implicated in functioning as vectors for transmitting plant diseases. In general, reports on pathogen transmission by spiders are very scant. Even though a few reports of spider-mediated dispersal of pathogens in cropping systems are available, these have never been demonstrated in disease pathogens infecting leguminous crops. However, recent reports on arthropodous predators revealed that spiders showed to rather influence the behaviour of some leaf-feeding herbivorous insects. Tholt *et al.* [31] reported the transmission of the Wheat Dwarf Virus through leafhoppers (*Psammotettix alienus*) as vectors in barley host plants. The study showed that leafhoppers changed their feeding behaviours in the presence of predator-spider risks, whereby the infection rate of plant tissues by the virus was decreased only when spiders were present. But another disadvantage is that, by feeding on most of the insects that are mostly found in legume crop fields, the spiders may indirectly decrease the number and reproductive rate of insects required for pollination [32].

Possibly, the negative impacts of spider activities on leguminous crops could be that spider webs and egg sacs may facilitate the spreading of aphids, mites, and other arthropods carrying pathogenic diseases that may affect plants. Nevertheless, spiders are fourth only to the malarial mosquito as insect vectors of fatal infections in humans. But to date, only black widow spiders and the brown recluse spiders that are native to many populous areas in southern America have been implicated in the transmission of human affecting diseases [33]. In plants, consideration should be given to whether small egg sacs of spiders that may get ingested could result in illnesses or deaths in humans and animals consuming infested crops, and if any such case would encourage people to stop buying pest-ridden crops. This also implies that farmers will lose out on their profits and subsequently their businesses. Furthermore, some field workers harvesting legumes may accidentally get bitten and poisoned.

PREVENTION AND CONTROL OF SPIDER MANIFESTATIONS

Currently, there are no insecticides that are effectively used to control and prevent the infestations of agricultural fields by spiders. Therefore, to prevent the manifestations of spiders in agroecosystems, foliar application of any insecticides that kill insect pests that spiders feed on must be carried out. Once the agricultural land does not have insects or pests, spiders may be reduced due to starvation.

There will then be fewer spider populations than there would have been if the pests had not been removed from the system. Most of the spider species do not feed on plants, so migration to other cropland or deaths because of starvation will take place. Although spiders are resistant to insecticides, different combinations of insecticide applications may succeed in decreasing the number of species found in the agroecosystem. The different combinations tested may mostly be effective in preventing severe future spider manifestations [34]. Furthermore, a method of biologically controlling the manifestations of spiders is called landscape diversity, wherein the structures and the quality composition of the landscape influence the activities of the spiders, the enemies, and the crops.

Some landscapes with non-crop habitats and abundant crops tend to have a high level of biocontrol activities, especially for legumes, such as soybeans, chickpeas, and cowpeas. Intentionally planting crops in these favourable landscapes improves the biological control of spiders and thus pest infestations [35]. On the other hand, spider manifestations in agroecosystems can be partially chemically controlled through the treatment of the area with residual liquid insecticides. Wettable powder (WP or WSP), microencapsulate (ME or CS) and suspended concentrate (SC) serve as some of the formulations that are often applied to provide durable or long-term protection against invasive or rapidly multiplying spider species. These formulations are applied with an aerosol or fan spray from a pressurized sprayer to avoid contamination of plant surfaces because the plants serve as sources of food [36].

Various insects show resistance to insecticides, especially after foliar application. Like many other insects, spiders have also gained genetic resistance that allows them to survive after pesticide exposure. These may possibly be because of natural responses or the resistance that has emanated from gene pool contamination from the cultivated genetically engineered crops [37, 38]. Under closed-off environments such as the greenhouse, the physical control of spider manifestations includes using pest-proofing measures that prevent spiders from entering buildings. Measures include repairing the screens, roofing, and sheathing on the exterior. Sticky traps or sticky monitors are also placed at various points within the greenhouse to detect spider activity and trap those that get attached to the sticky parts [39, 40]. An advanced tool that has a telescopic handle can also be used for the removal of spider webs; a regular feather duster can also be effective. Osakabe [41] demonstrated the use of ultraviolet-B radiation to prevent and control spider mites and other associated arthropods under greenhouse conditions. However, in large-scale agricultural lands that have been infested by arthropodous spiders, the application of the above-mentioned mechanical control methods may be a tedious or very long and tiring process [42].

CONCLUSION

Arthropodous spiders may be small organisms in size, but they are very beneficial to the agroecosystems that they live in. However, more research still needs to be done on their distribution patterns, diversity, and management so that they can be formally known and understood like many other insect pests. Such insights, including further discoveries, will contribute to the farmers' use of relevant beneficial spiders to their advantage and increase their profits, as well as help feed the entire population without any worries concerning spiders serving as problematic pests.

CONSENT FOR PUBLICATION

Not applicable.

CONFLICT OF INTEREST

The authors declare no conflict of interest, financial or otherwise.

ACKNOWLEDGEMENTS

The authors thank Dr. P Mangena for his motivation and guidance.

REFERENCES

[1]   Ubick, D.; Paquin, P.; Cushing, P.E.; Roth, V. *Spiders of North America: An identification manual*; American Arachnological Society: USA, **1999**, p. 377.

[2]   Sajid, M.; Zahid, M.; Shah, M.; Rasool, M.; Ullah, I.; Ahmad, R.; Habibullah, P.; Majeed, N. Identification of the orb weaving spider *(Araneae: Araneidae)* fauna of Dir Lower (Pakistan) through DNA barcoding. *J. Anim. Plant Sci.,* **2021**, *31*(4), 1197-1207.

[3]   Sebastian, S.; Gautam, A. Arthropod morphology. In: *Encyclopedia of Animal Cognition and Behaviour*; Vonk, J.; Shackelford, T.K., Eds.; Springer Nature Switzerland AG, **2021**. [http://dx.doi.org/10.1007/978-3-319-47829-6_823-1]

[4]   Isbister, G.K. Spider mythology across the world. *West. J. Med.,* **2001**, *175*(2), 86-87. [http://dx.doi.org/10.1136/ewjm.175.2.86] [PMID: 11483545]

[5]   Mammola, S.; Michalik, P.; Hebets, E.A.; Isaia, M. Record breaking achievements by spiders and the scientists who study them. *PeerJ,* **2017**, *5*, e3972. [http://dx.doi.org/10.7717/peerj.3972] [PMID: 29104823]

[6]   Gogoi, J.; Ningthoujam, K. Arthropod biodiversity in agricultural, horticultural and silvicultural ecosystems with special reference to spiders (Araneae) in Mid-hills of Meghalaya, India. *Res Square,* **2021**, *1*, 1-18. [http://dx.doi.org/10.21203/rs.3.rs-640064/v1]

[7]   Rypstra, A.L. Building a better insect trap; An experimental investigation of prey capture in a variety of spider webs. *Oecologia,* **1982**, *52*(1), 31-36. [http://dx.doi.org/10.1007/BF00349008] [PMID: 28310105]

[8]   Boutry, C.; Blamires, S.J. Plasticity in spider webs and silk: an overview of current evidence. In: *Insects and other terrestrial arthropods: Biology, Chemistry and Behaviour*; Santerre, M., Ed.; Nova

Science Publishers: USA, **2013**; p. 2.

[9]   Blamires, S.J. Spider webs as extended phenotypes. In: *Insects and Other Terrestrial Arthropods: Biology, Chemistry and Behaviour*; Santerre, M., Ed.; Nova Science Publishers: USA, **2013**; pp. 48-58.

[10]   Haupt, J. Taxonomy of Spiders. *Toxin Rev.,* **2005**, *24*(3-4), 249-256.
        [http://dx.doi.org/10.1080/07313830500236101]

[11]   Savory, T.H. *The biology of spiders*; Sidgwick and Jackson LTD: London, **1928**, pp. 116-133.
        [http://dx.doi.org/10.5962/bhl.title.82435]

[12]   de Almeida, A.M.R.; Specht, C.D. *Tropical biodiversity: Why Should We Care?*; Frontiers Media: Lausanne, **2019**.
        [http://dx.doi.org/10.3389/978-2-88945-878-3]

[13]   Vasconcellos-Neto, J.; Messas, Y.F.; Souza, H.S.; Villanueva-Bonila, G.A.; Romero, G.Q. Spider–plant interactions: An ecological approach. In: *Behaviour and Ecology of Spiders*; Viera, C.; Gonzaga, M.O., Eds.; Springer International Publishing AG, **2017**; pp. 165-205.
        [http://dx.doi.org/10.1007/978-3-319-65717-2_7]

[14]   Vollrath, F.; Selden, P. The role of behavior in the evolution of spiders, silks, and webs. *Annu. Rev. Ecol. Evol. Syst.,* **2007**, *38*(1), 819-846.
        [http://dx.doi.org/10.1146/annurev.ecolsys.37.091305.110221]

[15]   Riechert, S.E. Thoughts on the ecological significance of spiders. *Bioscience,* **1974**, *24*(6), 352-356.
        [http://dx.doi.org/10.2307/1296741]

[16]   Oberg, S. *Spiders in the agricultural landscape diversity, recolonisation, and body condition*; Doctoral Thesis, Swedish University of Agricultural Sciences, **2007**.

[17]   Ayoub, N.A.; Hayashi, C.Y. Spiders (Araneae). In: *The Time Tree of life*; Hedges, S.B.; Kumar, S., Eds.; Oxford University Press: London, **2009**; pp. 255-259.

[18]   Cranshaw, W.; Redak, R. *Bugs Rule! An Introduction to the World of Insects*; Princeton University Press: United Kingdom, **2013**, pp. 75-95.

[19]   Hadley, D. The life cycle of a spider. **2020**. Available from: https://www.thoughtco.com/the-spide--life-cycle-1968557 (Date accessed: 06/2022).

[20]   Rao, D. Habitat selection and dispersal. In: *Behaviour and Ecology of Spiders*; Viera, C.; Gonzaga, M.O., Eds.; Springer International Publishing AG, **2017**; pp. 85-104.
        [http://dx.doi.org/10.1007/978-3-319-65717-2_4]

[21]   Uetz, G.W.; Halaj, J.; Cady, A.B. Guild structure of spiders in major crops. *J. Arachnol.,* **1999**, *27*, 270-280.

[22]   Glime, J.M.; Lissner, J. Arthropods: Arachnida – Spider habitats. In: *Bryophyte Ecology, Volume 7---1*; Glime, J.M., Ed.; Michigan Technological University and the International Association of Bryologists: USA, **2013**; pp. 7-47.

[23]   Rypstra, A.L.; Carter, P.E. The web-spider community of soybean agroecosystems in Southwestern Ohio. *J. Arachnol.,* **1995**, *23*, 135-14.

[24]   Pearce, S.; Hebron, W.M.; Raven, R.J.; Zalucki, M.P.; Hassan, E. Spider fauna of soybean crops in south-east Queensland and their potential as predators of *Helicoverpa spp. (Lepidoptera: Noctuidae). Aust. J. Entomol.,* **2004**, *43*(1), 57-65.
        [http://dx.doi.org/10.1111/j.1440-6055.2003.00378.x]

[25]   Meena, L.K.; Sharma, A.N. Biodiversity, guild structure and vertical stratification of spiders in soybean ecosystem. *J. Entomol. Zool. Stud.,* **2019**, *7*(6), 1002-1004.

[26]   Rahmani, F.; Banan Khojasteh, S.M.; Ebrahimi Bakhtavar, H.; Rahmani, F.; Shahsavari Nia, K.; Faridaalaee, G. +Poisonous spiders: Bites, symptoms and treatment; an educational review.

Emergency, **2014**, *2*(2), 54-58.
[PMID: 26495347]

[27]   Michalko, R.; Pekár, S. Different hunting strategies of generalist predators result in functional differences. *Oecologia,* **2016**, *181*(4), 1187-1197.
[http://dx.doi.org/10.1007/s00442-016-3631-4] [PMID: 27098662]

[28]   Higgins, L.E.; Buskirk, R.E. Spider-web kleptoparasites as a model for studying producer-consumer interactions. *Behav. Ecol.,* **1998**, *9*(4), 384-387.
[http://dx.doi.org/10.1093/beheco/9.4.384]

[29]   Sunderland, K.; Samu, F. Effects of agricultural diversification on the abundance, distribution, and pest control potential of spiders: A review. *Entomol. Exp. Appl.,* **2000**, *95*(1), 1-13.
[http://dx.doi.org/10.1046/j.1570-7458.2000.00635.x]

[30]   Michalko, R.; Pekár, S.; Dul'a, M.; Entling, M.H. Global patterns in the biocontrol efficacy of spiders: A meta☐analysis. *Glob. Ecol. Biogeogr.,* **2019**, *28*(9), 1366-1378.
[http://dx.doi.org/10.1111/geb.12927]

[31]   Tholt, G.; Kis, A.; Medzihradszky, A.; Szita, É.; Tóth, Z.; Havelda, Z.; Samu, F. Could vectors' fear of predators reduce the spread of plant diseases? *Sci. Rep.,* **2018**, *8*(1), 8705.
[http://dx.doi.org/10.1038/s41598-018-27103-y] [PMID: 29880845]

[32]   Romero, G.Q.; Antiqueira, P.A.P.; Koricheva, J. A meta-analysis of predation risk effects on pollinator behaviour. *PLoS One,* **2011**, *6*(6), e20689.
[http://dx.doi.org/10.1371/journal.pone.0020689] [PMID: 21695187]

[33]   Isbister, G.K.; Framenau, V.W. Australian wolf spider bites (Lycosidae): Clinical effects and influence of species on bite circumstances. *J. Toxicol. Clin. Toxicol.,* **2004**, *42*(2), 153-161.
[http://dx.doi.org/10.1081/CLT-120030941] [PMID: 15214620]

[34]   Ishaaya, I.; Kontsedalov, S.; Horowitz, A.R. Emamectin, a novel insecticide for controlling field crop pests. *Pest Manag. Sci.,* **2002**, *58*(11), 1091-1095.
[http://dx.doi.org/10.1002/ps.535] [PMID: 12449526]

[35]   Schmidt, M.H.; Roschewitz, I.; Thies, C.; Tscharntke, T. Differential effects of landscape and management on diversity and density of ground-dwelling farmland spiders. *J. Appl. Ecol.,* **2005**, *42*(2), 281-287.
[http://dx.doi.org/10.1111/j.1365-2664.2005.01014.x]

[36]   Hazra, D.K.; Purkait, A. Role of pesticide formulations for sustainable crop protection and environment management: A review. *J. Pharmacogn. Phytochem.,* **2019**, *8*(2), 686-692.

[37]   Hazra, D.K.; Karmakar, R.; Rajlakshmi, P.; Bhattacharya, S.; Mondal, S. Recent advances in pesticide formulations for eco-friendly and sustainable vegetable pest management: A review. *Arch. Agric. Enviro. Sci.,* **2017**, *2*(3), 232-237.

[38]   Gatehouse, A.M.R.; Ferry, N.; Edwards, M.G.; Bell, H.A. Insect-resistant biotech crops and their impacts on beneficial arthropods. *Philos. Trans. R. Soc. Lond. B Biol. Sci.,* **2011**, *366*(1569), 1438-1452.
[http://dx.doi.org/10.1098/rstb.2010.0330] [PMID: 21444317]

[39]   Weintraub, P.G.; Berlinger, M.J. Physical control in greenhouses and field crops. In: *Insect Pest Management*; Horowitz, A.R.; Ishaaya, I., Eds.; Springer: Berlin, Heidelberg, **2004**; pp. 301-318.
[http://dx.doi.org/10.1007/978-3-662-07913-3_12]

[40]   Albajes, R.; Gullino, M.L.; van Lenteren, J.C.; Elad, Y. *Integrated pest and disease management in greenhouse crops*; Kluwer Academic Publishers: Dordrecht, **2002**, pp. 1-13.

[41]   Osakabe, M. Biological impact of ultraviolet-B radiation on spider mites and its application in integrated pest management. *Appl. Entomol. Zool.,* **2021**, *56*(2), 139-155.
[http://dx.doi.org/10.1007/s13355-020-00719-1]

[42]   Michalko, R.; Pekár, S.; Entling, M.H. An updated perspective on spiders as generalist predators in biological control. *Oecologia,* **2019,** *189*(1), 21-36.
[http://dx.doi.org/10.1007/s00442-018-4313-1] [PMID: 30535723]

CHAPTER 7

Role of Climate-Driven Factors on Bean Leaf Beetle, Corn Earworm and Stinkbug Populations, Control and their Effects on Soybean Growth and Productivity

Arinao Mukatuni[1,*]

[1] *Department of Chemical Sciences, Faculty of Science, University of Johannesburg, Doornfontein Campus, P. O. Box 17011, Johannesburg 2028, South Africa*

Abstract: Soybean is a crucial crop that is recognised globally for its high-value protein, vitamins, carbohydrates, fibre, and oils. However, the production of soybeans is frequently influenced by biotic stress factors such as bean leaf beetles (*Cerotoma trifurcate*), grasshoppers (*Schistocerca americana*), corn earworms (*Helicoverpa zea*) and stinkbugs (*Halyomorpha halys*). However, these insect pests were discovered to be both beneficial and harmful to crop growth and productivity, particularly, in soybeans. According to the literature, the rise in temperature causes an increase in insect pest populations, thereby severely influencing the growth, and yield quality of many crops. Less precipitation also contributes to drought stress, and plants undergoing water-deficit stress produce fewer secondary metabolites rendering them vulnerable to attacks by these insects. Similar effects were also revealed due to the rise in atmospheric CO_2 levels that led to the adverse weather effects that caused enhanced reproduction and spread of pest insects. This chapter, therefore, explores the role of climate change-induced factors, such as temperature, precipitation patterns and rising atmospheric CO_2 on insects' distribution, and reproductive patterns, as well as their subsequent influence on crop growth and productivity in soybeans. The review also briefly discusses the chemical, biological and biotechnological approaches of insect pest control that have been employed effectively to combat losses of crop production. Side effects, cost effectiveness and the ability of new biotechnological methods to target specific pests are also discussed in this chapter.

Keywords: Biotic stress, Climate change, Chemical control, Biological control, Insect pests, Genetic engineering, Soybean.

---

[*] **Corresponding author Arinao Mukatuni:** Department of Chemical Sciences, Faculty of Science, University of Johannesburg, Doornfontein Campus, P. O. Box 17011, Johannesburg 2028, South Africa; Tel: +2772-030-8129; E-mail: arinaom2@gmail.com

Phetole Mangena & Sifau A. Adejumo (Eds.)

INTRODUCTION

Soybean, *Glycine max* (L.) Merr., is a grain legume known to be an important agricultural crop globally and economically. This crop is grown worldwide, however, production is affected by various factors like abiotic, and biotic stress such as drought, feverish temperatures and insect pests. There are over 700 species of plant feeding insects for soybean but, the most notable damage is demonstrated by only eight species of insects found in the United States (US) [1]. These entails the velvetbean caterpillar, *Anticarsia gemmatalis*; the soybean looper, *Pseudoplusia includens*; the green cloverworm, *Plathypena scabra*; the Mexican bean beetle, *Epilachna varivestis*; the bean leaf beetle, *Cerotoma trifurcata*; the green stink bug, *Acrosternum hilare*; and the corn earworm, *Helicoverpa zea* [1, 2]. The bean leaf beetle is a native species in the eastern part of the US and a major pest in all soybean-growing areas across the globe [2].

The larvae of this beetle are said to feed on the roots, root hairs, and nodules of soybean, while the adults defoliate the leaves and feed on the external pod tissues [3]. *Helicoverpa zea* is an economically significant insect and said to be dominant also in the United States of America. This pest largely attacks soybean plants, affecting mainly the leaves during vegetative stages. The size of the *Helicoverpa zea* caterpillar, the development stage of the plant, the time of damage it inflicts, and the plant's ability to recover are some of the factors that can affect the yield of soybeans [4]. However, to compensate for the damage done to the reproductive tissues, soybeans can produce more pods or increase their seed's weight [3, 4]. Furthermore, the harmful insects known as the stinkbugs are also part of the growing problem of soybeans, and they feed on the pods while causing severe damage to the developing seeds [5].

Therefore, this chapter explores the role that insect pests play in soybean fields and examine the influence of climate change-induced factors such as temperature, precipitation patterns and rising atmospheric CO_2 on the insects' distribution, reproductive patterns, as well as their subsequent influence on crop growth and productivity in soybean. The review also briefly discusses the chemical, biological and biotechnological approaches of insect pest control that have been employed effectively to combat losses of crop production. Side effects, cost effectiveness and the ability of new biotechnological methods to target specific pests are also discussed in this chapter.

BIOLOGY OF BEAN LEAF BEETLE, CORN EARWORM AND STINKBUG INSECT PESTS

Bean Leaf Beetle (*Cerotoma trifurcata*, Family Chrysomelidae)

Cerotoma trifurcate also known as the bean leaf beetle (BLB) belongs to the Chrysomelidae family, order Coleoptera [6]. Bean leaf beetle are classified according to their appearance in colour. The colour of adult beetles differs from red to orange, as well as light yellow as exemplified in Fig. (**1**). The wing covers (elytra) of the adult BLB are soft and beige when first spotted. There are four squarish black markings on the underside of the wings, but these can be from few to nothing at all, and they are usually rimmed with a black margin. Black frons (faces) are typically found on female beetles, while tan frons are typically found on male ones [7]. The first tarsal segment of a male beetle also has a covering of dense setae (hairs), which are believed to be part of the mating process, this feature is not there in female beetles [7, 8]. The overall life span of this insect is usually 1 to 2 months [9].

Fig. (1). Examples of adult bean leaf beetles with different colours (**A-C**) [7], corn earworm *Helicoverpa zea* (**D**) [12] insect pests and adult brown marmorated stinkbug (**E**) [15].

Corn Earworm (*Helicoverpa zea*, Family Noctuidae)

Helicoverpa zea (Lepidoptera; Noctuidae) is known by many different names, including American cotton bollworm, corn earworm, cotton bollworm, bollworm, tomato fruitworm, soybean podworm, and sorghum headworm [10]. This insect pest is a highly polyphagous insect, and it is native to the United States of America (USA). Corn earworm is a major pest for a variety of crops, including corn, hemp, and soybean. This insect is nocturnal which means it inflicts more crop damage during the night [10, 11]. The corn earworm is also identified to be an economically significant agricultural pest found throughout the South and North American region [10]. The adults of corn earworms are generally around 20 to 25 mm long and have a wingspan of around 40 to 45 mm, identified by brown colour (Fig. **1D**) (females) and small brownish green colour which is for males [9]. They have small spots on their forewings and their underwings are distinguished by marginal bands that are dark on the outer part and brown disc-shaped spots [10]. Eggs are laid singly and after being laid they range from white to yellow near larval hatching the pupae are reddish brown and they are 20 mm long [9].

The life cycle of this insect usually takes about 30 days at 25°C to complete [11, 12]. The eggs are laid on the stigmas of female corn flowers, which are referred to as corn silk and this is where the newly hatched larvae begin feeding [11]. During their lifetime, female corn earworms can lay up to 3,000 eggs in a lifetime and up to 35 eggs in a day. Once they reach their larval stage in 3 to 4 days, they mature leave the feeding site, and go into the ground where they pupate. Additionally, adult corn earworm can develop and become active in 10 to 14 days [10].

Stinkbug (*Halyomorpha halys*, Family Pentatomidae)

Halyomorpha halys is a member of the Pentatomidae family, also known as shield or stinkbug [13]. The stinkbug is commonly known as the brown marmorated stinkbug because it is about 12 to 17 mm long and has a marbled brown color (Fig. **1E**) [14]. Stinkbugs come in a variety of shapes and sizes, but most of them have round or oval (sometimes shield-shaped) bodies, well-developed and often triangle-shaped scutellum, piercing-sucking mouthparts, and five-segmented antennae [14]. They are fairly large and emit foul odors from glands on the ventral side of the thorax [13]. Adult *Halyomorpha halys* is distinguished from native stinkbug species in the US by white and black banding on their antenna and abdominal edges. Adult males and females are distinguished by claspers located on the last ventral abdominal segment [15]. This insect produces one to two generations per year, and during development, it undergoes incomplete metamorphosis, with very small differences between nymphs and adults [13].

Nymphs develop in five stages whereby with each molt, they grow bigger and resemble fully developed insects. Nymphs have a length of 2.4 to 12.0 mm, and their body shape changes from elliptical to pear-shaped as they develop, while their abdomen colour changes from yellow-orange with black markings to brown as they progress from the 1st to 5th instars [14]. The development period for these adult corn earworm larvae is 32 to 35 days at 30°C starting from an egg to being an adult and their temperature preference range is from 14 to 35°C [15].

EFFECT OF CLIMATE-RALATED FACTORS ON ECOLOGY AND SPECIES DISTRIBUTION

Bean Leaf Beetle (*Cerotoma trifurcata*, Family Chrysomelidae)

The bean leaf beetle is primarily found in the Western and Eastern United States [9]. Chrysomelidae, one of the richest and most diverse families of herbivorous insects, has long been used to study the evolution of host specification in phytophagous insects [6]. They have over 50 000 species and over 2000 genera and are found all over the world [16]. They are commonly found in fields, particularly those that grow soybeans, pumpkins, cucumbers, green beans, and other grain legumes [9]. This insect is considered an occasional soybean pest; however, its pest status, abundance, and significance have increased in recent years, owing primarily to the rapid increase in soybean acreage. Environmental conditions, overwintering success, cultural practices, natural enemies, cultivar planted, and synchronization of soybean and beetle development are all factors that can influence beetle population growth [17]. This family contains both beneficial and harmful beetle species. For example, many Chrysomelidae beetles are commercial agricultural pests and are thus considered important economic species [6]. Bean leaf beetles can survive freezing temperatures and overwinter as adults [18].

Corn Earworm (*Helicoverpa zea*, Family Noctuidae)

Except for northern Canada and Alaska, corn earworm can be found throughout South and North America [19]. The *Helicoverpa zea* is active all year in subtropical and tropical climates, but it is usually restricted to the summer season as the latitude increases [12]. This species overwinters in North America during the pupal stage and adults emerge in the spring when there is a rise in temperatures [19]. This insect has been observed to survive as far north under 40° north latitude conditions, which corresponds to Kansas, Ohio, Virginia, and southern New Jersey, depending on the severity of the winter weather. It is, however, widely dispersed and frequently spread from southern states to northern states and Canada [12]. Northern *Helicoverpa zea* populations exist solely as migrant populations from the south because the diapausing pupae can't stand winter temperatures north of 40° N latitude. As such, adult *Helicoverpa zea* overwinters in the southern states and then migrate north as the season progresses. Depending on the weather and location, different areas may have immigrant or overwintering populations [12, 19].

Stinkbug (*Halyomorpha halys*, Family Pentatomidae)

Halyomorpha halys, also known as brown marmorated stinkbug, is a polyphagous stinkbug that is native to China, Japan, Korea, and Taiwan [20]. Stinkbugs are the third largest family of Heteroptera, with over 4700 species (more than 10% of the entire order) organized into 900 genera and 8–11 subfamilies [21]. This insect pest is invasive, and it is believed that the economic impact of its infestations increases as the population grows. For instance, since 2012, the *Halyomorpha halys* have been found in Ohio soybean fields even at the economic level with other stinkbug species [14]. There are speculations that this species was introduced to the US through a single introduction from Beijing, China [15]. *Halyomorpha halys* has spread across most of the United States, becoming established in Canada and some European countries, causing significant threats to agricultural production since the early 2000s [13]. Moderate stinkbugs that overwinter as adults can survive the winter in protected areas such as under-leaf litter and debris. The adults of this species also spend their winter in buildings or under loose tree bark [14]. Although, most sting bugs overwinter as adults some overwinter in different developmental stages, for example, a few species, including *Picromerus bidens* and *Apateticus cynicus*, overwinter as eggs, while others, including *Carbula humerigera* and *Pentatoma rufipes*, overwinter as nymphs [21].

HOW DOES CLIMATE-DRIVEN FACTORS AFFECT INSECT PESTS?

Climate change is a global concern that is thought to have negative and positive effects on the agricultural sector, with the negative effects outweighing the positive ones. The effects include increase in temperatures, distorted precipitation patterns, and rising CO_2 concentrations which all have a significant impact on living organisms, from species to ecosystem levels [22]. Regional climate change may cause local population declines or extinctions of insect populations [23]. This implies that pest populations are also expected to fluctuate in response to climate change, which might either decrease or increase the number of insect pests found in a habitat [24]. Like other living organisms affected by fluctuating climate conditions, insect species with strong defence mechanisms are more likely to survive changes in climatic conditions, while others will become extinct.

Temperature

Temperature has a variety of effects on insect species. Since some insects are cold-blooded (poikilothermic), temperature has a strong influence on their growth, development, and multiplication [24]. As such, this will usually result in an alteration of the insect dynamics. Species that cannot adapt and thrive in higher temperatures have a difficult time maintaining their population, whereas others thrive and multiply rapidly. Warmer temperatures are beneficial to insects because they are exothermic [25]. It was discovered that a 2°C temperature increase causes insects to experience one to five additional life cycles per season, and if mortality per generation does not change, the insect population will possibly increase under global warming. Winter temperature increases may aid in the survival of some pests. This is because less winter mortality of insects due to warmer winter temperatures may play a role in increasing the insect population [26].

Environmental factors associated with rising temperatures that can affect the development and maintenance of crops are the availability of water, the duration and intensity of sunlight, and strong winds. Rising temperatures can additionally cause atmospheric water demand to increase, this can result in higher water pressure deficits, reduce soil moisture, and eventually affect crop yield [22]. Temperature influences metabolism, metamorphosis, mobility, host availability, and other factors that influence the possibility of change in insect pest population and dynamics [24]. It also regulates insect physiology and metabolism; thus, increasing physiological activities and, consequently, metabolic rates [22]. Insects must consume more food to survive, and insect herbivores are expected to consume more and grow faster, increasing the insect population since reproduction would also be faster [27]. However, crop damage will also increase because of this expansion.

Relative Humidity/Precipitation Patterns

Events caused by climate change are taking place daily even though some effects do not manifest immediately. The distribution and frequency of rainfall have an impact on pest occurrence, both directly and indirectly through changes in humidity levels [28]. As a result, many global climate models predict that because of climate change, rainfall patterns will change, and storms will become more severe [29]. The types of rainfall patterns that cause droughts and floods have a direct impact on insect populations, especially those that overwinter within the soil. This places insects' lives in danger and, at the very least, messes up with their diapause [22]. Furthermore, insect eggs and larvae, as well as tiny pests such as aphids, mites, and jassids can be wiped away by heavy rains and flooding [28].

Large wind snaps or windthrow is also a result of altered precipitation patterns. It usually causes bark beetle outbreaks by giving an abundance of the breeding substrate in the form of storm-felled or broken trees, which are not able to induce tree defences the same as standing healthy trees [30]. This then explains the rise in the death of trees because of bark beetles. Drought affects herbivorous insects in many ways. Many pest species prefer warm environments to thrive and as a result, moisture stress makes crops more vulnerable, either directly or indirectly to pest damage during the initial stages of crop growth [31]. Drought-stressed plants are more vulnerable to insect attack because they produce fewer secondary metabolites with a defence action [22].

Rising Atmospheric CO_2

Increases in atmospheric carbon dioxide (CO_2) have the potential to change plant chemistry, affecting crop susceptibility to insect herbivores [32]. The effects of increased atmospheric CO_2 levels are not always negative; they can also be beneficial to plants. This is corroborated by Dermody *et al.* [33] who claimed that an increase in atmospheric CO_2 due to the burning of fossil fuels and changes in land usage accelerates the rate of photosynthesis in plants, leading to an increase in plant productivity. The study noted that the plants associated with nitrogen-fixing symbionts such as alfalfa, soybeans, and lupine, are more likely to benefit from the increased CO_2 resource under favourable environmental conditions. Plant species may also react to increased CO_2 levels in the atmosphere by changing the quality of their leaves. The nutritional quality of leaves declined significantly in experimentally enriched CO_2 atmosphere due to nitrogen dilution by 10-30% [34].

Lower foliar nitrogen content caused by elevated CO_2 causes herbivores to consume up to 40% more food, however, reduced leaf nitrogen and compensatory feeding are less likely in legumes, which may overcome nitrogen limitation through increased nodulation and N fixation. Although many invasive insects can adapt to a variety of environmental conditions, rising global temperatures and CO_2 levels could allow the spread of non-native pests, especially in the northern and mid-latitudes [33]. As a result, non-native insects will predominate, endangering native insect populations and diversity. Cotton bollworm (*Helicoverpa armigera* Hübner) is considered a critical pest due to crop losses totaling $7 billion per year and its ability to spread to new areas. As a result, many studies were conducted to determine which crops are vulnerable to *Helicoverpa armigera* at predicted atmospheric CO_2 concentrations, and the reports suggested four plant families, namely the Fabaceae, Malvaceae, Poaceae, and Solanaceae [32].

IMPACT OF INSECT PESTS ON SOYBEAN GROWTH AND PRODUCTIVITY

Insect pests among other factors are known to reduce the growth and productivity of soybeans and other leguminous and non-leguminous crops [1]. However, soybean plants are highly susceptible to insects throughout their growth stages which makes them easy hosts for different insect pests. There are various insects that may be found in fields where soybean is grown, and their availability varies according to seasons. Asala *et al* [35]. reported that about 20% of the production of soybeans is lost due to damage caused by pests. Out of the 29 species of insects that can cause major damage to soybean plants, nine of them are vectors of plant viruses that use green living plants as their habitats. About 70% of the insects that can cause this damage belong to one of the orders Homoptera, and amongst them, amphids are the most important species of insects, while grasshoppers (Orthoptera) are of lesser importance [35]. The bean leaf beetle, grasshoppers, corn earworm, and stinkbug are some of the insect pests that negatively affect soybean growth and productivity. The bean leaf beetle is known to cause the bean pod mottle virus which leads to the discoloration of the soybean plant as well as reducing the yield of the beans [36]. These insect pests target crops that are still in early growth stages and not fully developed. When the BLB heavily accumulates on a soybean crop, it causes extreme damage to the seedling, leaves, and pods (Fig. **2**).

Grasshoppers feed on soybean leaves and pods resulting in holes and damaged seeds. The rate at which grasshoppers consume leaves varies with seed cultivars, however, they mostly prefer soft leaves because they are easy to chew [39]. Once the leaves are damaged, the plant's ability to photosynthesise is lowered and growth becomes slower resulting in low productivity. Corn earworms usually feed on soybean foliage, pods, seeds, and flowers [37]. The large corn earworm larvae mainly attack the fully developed soybean seeds and lower the crop's metabolism. The effects of corn earworm feeding on soybeans result in reduced leaf surface area, number of seeds in each pod, and loss of yield. It has been reported that major insect pests that feed on soybeans are tobacco caterpillar, *Sopdoptera litura* (Fab.) and green semilooper, *Chrysodeixis acuta* which are defoliators [40]. These major insect pests mainly destroy the plant at its vegetative stage and in the worst instant, they defoliate the whole crop and reduce production. They cause a reduction in the leaf's photosynthetic capability, senescence of the leaves, phenological reduction, and other effects as indicated in Fig. (**2**) [1].

Fig. (2). Bean leaf beetle soybean leaf (**A**), illustration of pod damage (**B**) [37], and a picture of a tobacco caterpillar feeding on soybean pod (**C**) [38].

According to Brahman *et al.* [40] leaf damage inflicted by these insects can be about 40% and pod damage about 50%. Soybean can only tolerate 40% of defoliation before flowering and 25% after flowering without affecting productivity and yield. The destruction of the pods by insects reduces productivity and growth in the early to middle reproductive growth stages, and then the plant cannot remunerate for the damage during these stages because defoliation is at 80% which is unbearable resulting in damaged seeds and reduced seed viability [1]. It is quite hard to assess the damage caused by insect pests because various factors are involved, like the genotype or cultivar, environmental conditions, seasons, and the farming method used. A study was done to compute the impact of insect pests on the growth of soybean varieties (TGx144-7t, TGx1835-10f, and TGx1485-1d) in the Guinean savanna (Nigeria) as presented in Table **1** [35]. The findings suggested that cultivar TGx1835-10f had a lower germination rate compared to other cultivars and pest symptoms for this specific cultivar were more extreme with low crop yield. These observations were based on the different seasons, humidity, and other environmental factors, illustrating the different impacts of these pests based on different environmental factors.

Table 1. Mean percentage germination, incidence, and severity of virus symptom expression on soybean grown [35].

Treatment	Plant Height (cm)	Stem Diameter (cm)	Yield (kg/ha)
TGX 1440-7t	37.7[a]	2.1[a]	15,400[a]
TGX 1835-10f	27.3[b]	2.2[a]	2,900[b]
TGX 1485-1d	29.4[b]	2.0[a]	13,500[a]
Abuja local variety	29.9[b]	2.0[a]	14,100[a]
Values in the same column with different superscripts differ significantly (P<0.05) at a 5% level of probability.			

CHEMICAL CONTROL

Chemical control of pests is the use of chemical pesticides to lessen or prevent plant damage brought by insects. This method can be used effectively; however, it has side effects, and they can be severe depending on the chemicals involved. For example, some insecticides when overused, worsen the pest problem by only killing insects resistant to that specific insecticide. Chemicals that are used tend to be costly and negatively affect the environment. Literature highlights the importance of establishing a balance between pesticide costs and their effectiveness in controlling pests and the fact that it is advisable to use another method if the chances of effective control are low and the pesticide cost is very high [41]. The effectiveness of the pesticides depends on the climatic conditions, the extent of the damage, and the crop genotype or cultivar. Singh and Patel [42] assessed the effectiveness of chemicals to control insects in soybean crops. They mainly focused on five insect pests, namely the sapsucker, girdle beetle, cricket, leaf miner and 1 foliage feeder.

The insects were assessed for pest control using Endosulfan 35 EC, Cypermethrin 25 EC and Monocrotophol 36 SC pesticide chemicals. The results indicated that Monocrotophol effectively controlled sap sucker insects. Cypermethrin and Endosulfan were less effective against these insects. However, Cypermethrin was more effective for controlling foliage feeders and field crickets. Similarly, Vieira *et al.* [41] reported that Piriproxifen can be used to control whiteflies in soybeans due to its ability to suppress metamorphosis and formation of adult insects that affect the crop. Soybean growers use this insecticide to control the whitefly, especially at low infestations where damages to the crop could not cause loss in revenue. Jardine [43] also reported the application of Pyrethroid against the bean leaf beetle (BLB), applied immediately after crop emergence particularly when BLB symptoms are observed.

Pyrethroid is a preferred insecticide for this insect pest because it can grant protection even for the pests that will arise at later stages of growth. Helicoverpa (NVP) is a bioinsecticide used to treat small larvae of corn earworm and tobacco budworm [44]. Nevertheless, this insecticide is not effective if it is applied when there is rainfall. Chemical control of corn earworm in soybean can be done using Carbamate and Pyrethroid insecticides [19]. It was discovered that these two are the fastest-acting insecticides against corn earworms. Adamu *et al.* [45] also reported that the insecticide Prophylactic can be used to control stinkbugs by spraying it on the affected crops to control bugs. This further improves plant metabolism, germination and photosynthetic capacity of the affected crop. Among the various pesticides that are available, each pest has specific pesticides that can effectively control the insect through different mechanisms and the end goal is to enhance the plant metabolism and increase growth and productivity.

BIOLOGICAL CONTROL

Chemical pesticides are the main tool used by farmers to combat insect pests; nevertheless, frequent application of chemical insecticides has led to the emergence of insecticide resistance in naturally occurring populations of the insect pest [46]. As a result, alternatives to chemical insecticides have been developed, one of which is biological control. Biopesticides are made from organic components like microbes, plants, and minerals. They are effective and preferred because they are less harmful to the environment than chemical insecticides [47]. Microbial pesticides, plant-incorporated protectants (PIPs), and biochemical pesticides are the three major types of biopesticides [48]. Bacterial biopesticides are the most widely used and least expensive type of microbial pesticide. *Bacillus thuringiensis* (*Bt*) is a *Bacillus* species that is widely used to control insects [47, 48].

This bacteria-based mechanism for controlling insects is grounded on its toxins which serve as insecticidal toxins killing the host insect. Various *Bt* commercial formulations were tested against the legume pod borer (LPB), which include Dipel® (derived from *B. thuringiensis* subsp. kurstaki) and Florbac® (from *B. thuringiensis* subsp. aizawai). These formulations proved to be highly effective against the LPB. Entomopathogenic fungi form part of microbial organisms used for pest management and are a promising microbial biopesticide with a variety of pathogenic mechanisms [49]. These are well-known as obligatory or facultative organisms that can kill sucking insect pests, such as aphids, thrips, mealy bugs, whiteflies, scale insects, mosquitoes, and all types of mites based on contact [49].

Another type of biopesticide is plant-incorporated protectants (PIPs), which are defined as pesticidal substances designed to be produced and used in a living plant or its products, as well as the genetic material required to produce such a pesticidal substance [50]. Unfortunately, despite their effectiveness at killing pests, PIPs can cause allergic reactions, skin and eye irritation, and toxicity as well as birth defects, cancer, and neurological and reproductive system disorders [51]. Lastly, biochemical pesticides are naturally occurring substances that use nontoxic mechanisms to control pests and they differ from conventional pesticides in terms of their source and mode of pest control. These pesticides use substances that interfere with mating, like sex pheromones, as well as different fragrances of plant extracts like vegetable oils that attract pests, trapping and killing them [47, 51].

BIOTECHNOLOGICAL CONTROL

RNA interference (RNAi), gene transformation, genome editing, and other biotechnological approaches are used to control insect pests. To give crop plants resistance to insect pests, crop gene transformation for insect resistance is used and it requires inserting a certain DNA segment or gene into the crop plants. One technique for managing insect pests is the *cry* gene transformation, which includes introducing specific DNA sequences or genes into agricultural plants via particle bombardment or transformation mediated by *Agrobacterium* [52]. The DNA segment that is introduced usually encodes an insecticidal protein. Since their introduction, genetically modified crops that produce insecticidal crystalline proteins (ICPs) from *B. thuringiensis* have been widely used globally in agriculture. ICPs encoded by *cry* genes are hazardous to insect orders, such as Lepidoptera, Diptera, Coleoptera, Hymenoptera, and Homoptera [53].

The crystal protein mode of action involves protoxin protein which gets dissolved in the larval midgut due to alkaline pH. The protein gets bisected by enzymes to an active toxin that diffuses through the peritrophic membrane covering the gut and binds to receptors present in the midgut epithelium [52]. As a result, the pest's gut becomes paralyzed, and it stops eating and ends up dead in about 2 to 3 days. RNA interference (RNAi) is the process of suppressing gene expression at a specific sequence level [54]. The goal of the RNAi approach is to control the development of harmful genes in the target pest system by double-stranded RNA expression. For instance, the use of dsRNAs in maize resulted in the reduction of harmful genes in the target pest system, increasing the mortality rate of stem borer and piercing insects which came about after gene knockdown was observed [55].

In another example, the expression of dsRNAs in soybeans resulted in a reduction in the mortality rate of the soybean pod borer larvae [52, 56].

Another approach is the genome editing technique, a process that involves deleting, replacing, or inserting DNA bases in a specific sequence of the genome to alter the function of the gene [52]. Two of the most commonly used genome editing tools are the transcriptional activator-like effector nucleases (TALENs) and the clustered regularly interspaced short palindromic repeats and their associated endonucleases (CRISPR/Cas9). These approaches are known to perform various tasks, such as altering the function of a gene by binding DNA bases [57]. CRISPR system is commonly used to perform genome editing utilizing invariable Cas9 protein, and TALENs are redesigned several times to alter the function of a gene [58]. As a result, CRISPR is more widely used and technically simple. For instance, CRISPR/Cas9 has been successfully used to knock out many insect genes, including *H. armigera* and *S. exigua* [54].

CONCLUSION

Changes involving climatic conditions are a worldwide concern, and agriculture is one of the most vulnerable sectors severely influenced by such fluctuations. Crop losses are being exacerbated by climate-driven factors such as rising temperatures, CO_2 levels, and changes in rainfall patterns. Insect pests have also been identified as one of the agricultural problems influenced by climate change effects that contribute to losses in crops such as soybeans. Soybeans like any other legumes are vulnerable to insect pest attacks, particularly, under stressful conditions such as drought and higher temperatures. However, stress induced by insect pests and exacerbated by climate-driven factors may be controlled using chemical, biological, and biotechnological approaches. Biological and biotechnological controls remain preferred methods of choice due to their effectiveness, affordable costs, and ability to only target specific insects.

LIST OF ABBREVIATIONS

BLB	Bean leaf beetle
Bt	Bacillus thuringiensis
Cas	CRISPR-associated endonuclease proteins
Cry	Crystal protein
CRISPR	Clusters of regularly interspaced short palindromic repeats
DNA	Deoxyribose nucleic acid
dsRNA	Double stranded RNA
GMO	Genetically Modified Organisms
ICPs	Insecticidal crystalline proteins

LPB	Legume pod borer
PIPs	Plant-incorporated protectants
RNA	Ribonucleic acid
RNAi	RNA interfererence
TALES	Transcription Activator-Like Effector Nucleases
US	United States

CONSENT FOR PUBLICATION

Not applicable.

CONFLICT OF INTEREST

The author declares no conflict of interest, financial or otherwise.

ACKNOWLEDGEMENT

Declared none.

REFERENCES

[1]   Rupe, J.; Luttrell, R.G. Effect of pests and diseases on soybean quality. In: *Soybeans*; Academic Press and AOCS Press, **2008**; pp. 93-116.
        [http://dx.doi.org/10.1016/B978-1-893997-64-6.50007-X]

[2]   Lam, W.K.F.; Pedigo, L.P. A predictive model for the survival of overwintering bean leaf beetles *(Coleoptera: Chrysomelidae). Environ. Entomol.,* **2000**, *29*(4), 800-806.
        [http://dx.doi.org/10.1603/0046-225X-29.4.800]

[3]   Kanyesigye, D.; Alibu, V.P.; Tay, W.T.; Nalela, P.; Paparu, P.; Olaboro, S.; Nkalubo, S.T.; Kayondo, I.S.; Silva, G.; Seal, S.E.; Otim, M.H. Population genetic structure of the bean leaf beetle *Ootheca mutabilis* (Coleoptera: Chrysomelidae) in Uganda. *Insects,* **2022**, *13*(6), 543.
        [http://dx.doi.org/10.3390/insects13060543] [PMID: 35735880]

[4]   Coelho, M.; Cook, D.R.; Catchot, A.L.; Gore, J.; Lourenção, A.L.; Baldin, E.L.L. Simulated corn earworm, Helicoverpa zea, injury in an indeterminate soybean cultivar at various Growth Stages under non-irrigated conditions in the Southern United States. *Agronomy,* **2020**, *10*(10), 1450.
        [http://dx.doi.org/10.3390/agronomy10101450]

[5]   Pezzini, D.T.; DiFonzo, C.D.; Finke, D.L.; Hunt, T.E.; Knodel, J.J.; Krupke, C.H.; McCornack, B.; Michel, A.P.; Philips, C.R.; Varenhorst, A.J.; Wright, R.J.; Koch, R.L. Community composition, abundance, and phenology of stink bugs *(Hemiptera: Pentatomidae)* in soybean in the North Central Region of the United States. *J. Econ. Entomol.,* **2019**, *112*(4), 1722-1731.
        [http://dx.doi.org/10.1093/jee/toz099] [PMID: 31038171]

[6]   Azad, M.W.A.; Naeem, M.; Bodlah, I.; Mohsin, A.U. New locality records of Chrysomelidae *(Coleoptera)* from Pothowar tract of the Punjab. *Asian. J. Agri. Biol.,* **2015**, *3*, 41-45.

[7]   Hadi, B.A.R.; Bradshaw, J.D.; Rice, M.E.; Hill, J.H. Bean leaf beetle *(Coleoptera: Chrysomelidae)* and bean pod mottle virus in soybean: biology, ecology, and management. *J. Integr. Pest Manag.,* **2012**, *3*(1), B1-B7.
        [http://dx.doi.org/10.1603/IPM11007]

[8] Hammack, L.; French, B.W. Sexual dimorphism of basitarsi in pest species of diabrotica and cerotoma *(Coleoptera: Chrysomelidae). Ann. Entomol. Soc. Am.,* **2007**, *100*(1), 59-63.
[http://dx.doi.org/10.1603/0013-8746(2007)100[59:SDOBIP]2.0.CO;2]

[9] CABI ISC (Invasive Species Compendium). Bean leaf beetle (*Cerotoma trifurcate*). **2020**. Available from: https://beetleidentifications.vom/bean-leaf-beetle/ (Accessed: 27 September 2022).

[10] Bragard, C.; Dehnen-Schmutz, K.; Di Serio, F.; Gonthier, P.; Jacques, M.A.; Jaques Miret, J.A.; Justesen, A.F.; Magnusson, C.S.; Milonas, P.; Navas-Cortes, J.A.; Parnell, S.; Potting, R.; Reignault, P.L.; Thulke, H.H.; Van der Werf, W.; Civera, A.V.; Yuen, J.; Zappalà, L.; Czwienczek, E.; Streissl, F.; MacLeod, A. Pest categorisation of *Helicoverpa zea. EFSA J.,* **2020**, *18*(7), e06177.
[http://dx.doi.org/10.2903/j.efsa.2020.6177] [PMID: 32665793]

[11] Bilbo, T.R.; Reay-Jones, F.P.F.; Reisig, D.D.; Greene, J.K.; Turnbull, M.W. Development, survival, and feeding behavior of *Helicoverpa* zea (Lepidoptera: Noctuidae) relative to Bt protein concentrations in corn ear tissues. *PLoS One,* **2019**, *14*(8), e0221343.
[http://dx.doi.org/10.1371/journal.pone.0221343] [PMID: 31425563]

[12] Capinera, J.L. *Corn earworm, Helicoverpa (= Heliothis) zea (Boddie)(Lepidoptera: Noctuidae).* Florida Cooperative Extension Service, Institute of Food and Agricultural Sciences; University of Florida, **2000**.

[13] Pajač Živković, I.; Skendžić, S.; Lemić, D. Rapid spread and first massive occurrence of *Halyomorpha halys (Stål, 1855)* in agricultural production in Croatia. *J. Cent. Eur. Agric.,* **2021**, *22*(3), 531-538.
[http://dx.doi.org/10.5513/JCEA01/22.3.3173]

[14] Koch, R.L.; Pezzini, D.T.; Michel, A.P.; Hunt, T.E. Identification, biology, impacts, and management of stink bugs *(Hemiptera: Heteroptera: Pentatomidae)* of soybean and corn in the Midwestern United States. *J. Integr. Pest Manag.,* **2017**, *8*(1)
[http://dx.doi.org/10.1093/jipm/pmx004]

[15] Rice, K.B.; Bergh, C.J.; Bergmann, E.J.; Biddinger, D.J.; Dieckhoff, C.; Dively, G.; Fraser, H.; Gariepy, T.; Hamilton, G.; Haye, T.; Herbert, A.; Hoelmer, K.; Hooks, C.R.; Jones, A.; Krawczyk, G.; Kuhar, T.; Martinson, H.; Mitchell, W.; Nielsen, A.L.; Pfeiffer, D.G.; Raupp, M.J.; Rodriguez-Saona, C.; Shearer, P.; Shrewsbury, P.; Venugopal, P.D.; Whalen, J.; Wiman, N.G.; Leskey, T.C.; Tooker, J.F. Biology, ecology, and management of brown marmorated stink bug *(Hemiptera: Pentatomidae). J. Integr. Pest Manag.,* **2014**, *5*(3), 1-13.
[http://dx.doi.org/10.1603/IPM14002]

[16] Hangay, G.; Zborowski, P.A. *Guide to the Beetles of Australia*; CSIRO publishing, **2010**, pp. 1-238.
[http://dx.doi.org/10.1071/9780643100121]

[17] Tiroesele, B.; Hunt, T.E.; Wright, R.; Foster, J.E. Population dynamics of bean leaf beetle, *Cerotoma trifurcata* (Coleoptera: Chrysomelidae) on edamame soybean plants in Nebraska. *Eur. J. Sustain. Dev.,* **2013**, *2*(1), 19-19.

[18] Lam, W.K.F.; Pedigo, L.P.; Hinz, P.N. Population dynamics of bean leaf beetles *(Coleoptera: Chrysomelidae)* in central Iowa. *Environ. Entomol.,* **2001**, *30*(3), 562-567.
[http://dx.doi.org/10.1603/0046-225X-30.3.562]

[19] Swenson, S.J.; Prischmann-Voldseth, D.A.; Musser, F.R. Corn earworms *(Lepidoptera: Noctuidae)* as pests of soybean. *J. Integr. Pest Manag.,* **2013**, *4*(2), 1-8.
[http://dx.doi.org/10.1603/IPM13008]

[20] Hemala, V.; Kment, P. First record of *Halyomorpha halys* and mass occurrence of *Nezara viridula* in Slovakia *(Hemiptera: Heteroptera: Pentatomidae). Plant Prot. Sci.,* **2017**, *53*(4), 247-253.
[http://dx.doi.org/10.17221/166/2016-PPS]

[21] Saulich, A.K.; Musolin, D.L. Diapause in the seasonal cycle of stink bugs *(Heteroptera, Pentatomidae)* from the Temperate Zone. *Entomol. Rev.,* **2012**, *92*(1), 1-26.
[http://dx.doi.org/10.1134/S0013873812010010]

[22] Skendžić, S.; Zovko, M.; Živković, I.P.; Lešić, V.; Lemić, D. The impact of climate change on agricultural insect pests. *Insects,* **2021**, *12*(5), 440.
[http://dx.doi.org/10.3390/insects12050440] [PMID: 34066138]

[23] Lehmann, P.; Ammunét, T.; Barton, M.; Battisti, A.; Eigenbrode, S.D.; Jepsen, J.U.; Kalinkat, G.; Neuvonen, S.; Niemelä, P.; Terblanche, J.S.; Økland, B.; Björkman, C. Complex responses of global insect pests to climate warming. *Front. Ecol. Environ.,* **2020**, *18*(3), 141-150.
[http://dx.doi.org/10.1002/fee.2160]

[24] Shrestha, S. Effects of climate change in agricultural insect pest. *Acta. Scientific. Agricul.,* **2019**, *3*(12), 74-80.
[http://dx.doi.org/10.31080/ASAG.2019.03.0727]

[25] Naeem-Ullah, U.; Ramzan, M.; Bokhari, S.H.M.; Saleem, A.; Qayyum, M.A.; Iqbal, N. Insect pests of cotton crop and management under climate change scenarios. In: *In Environment, climate, plant, and vegetation growth*; Fahad, S.; Saeed, S., Eds.; Springer: Cham, **2020**; pp. 367-396.

[26] Yadav, T.; Baloda, A.S.; Jakhar, B.; Yadav, A. Impact of climate changes on insect pest and integrated pest management: A review. *J. Entomol. Zool. Stud.,* **2019**, *7*, 1150-1156.

[27] Dukes, J.S.; Pontius, J.; Orwig, D.; Garnas, J.R.; Rodgers, V.L.; Brazee, N.; Cooke, B.; Theoharides, K.A.; Stange, E.E.; Harrington, R.; Ehrenfeld, J.; Gurevitch, J.; Lerdau, M.; Stinson, K.; Wick, R.; Ayres, M. Responses of insect pests, pathogens, and invasive plant species to climate change in the forests of northeastern North America: What can we predict?This article is one of a selection of papers from NE Forests 2100: A synthesis of climate change impacts on forests of the northeastern US and eastern canada. *Can. J. For. Res.,* **2009**, *39*(2), 231-248.
[http://dx.doi.org/10.1139/X08-171]

[28] Pathak, H.; Aggarwal, P.K.; Singh, S.D. *Climate change impact, adaptation, and mitigation in agriculture: Methodology for assessment and applications*; Indian Agricultural Research Institute: New Delhi, **2012**, p. 302.

[29] Trumble, J.T.; Butler, C.D. Climate change will exacerbate California's insect pest problems. *Calif. Agric.,* **2009**, *63*(2), 73-78.
[http://dx.doi.org/10.3733/ca.v063n02p73]

[30] Jactel, H.; Koricheva, J.; Castagneyrol, B. Responses of forest insect pests to climate change: Not so simple. *Curr. Opin. Insect Sci.,* **2019**, *35*, 103-108.
[http://dx.doi.org/10.1016/j.cois.2019.07.010] [PMID: 31454625]

[31] Yihdego, Y.; Salem, H.S.; Muhammed, H.H. Agricultural pest management policies during drought: Case studies in Australia and the state of Palestine. *Nat. Hazards Rev.,* **2019**, *20*(1), 05018010.
[http://dx.doi.org/10.1061/(ASCE)NH.1527-6996.0000312]

[32] Johnson, S.N.; Waterman, J.M.; Hall, C.R. Increased insect herbivore performance under elevated CO_2 is associated with lower plant defence signalling and minimal declines in nutritional quality. *Sci. Rep.,* **2020**, *10*(1), 14553.
[http://dx.doi.org/10.1038/s41598-020-70823-3] [PMID: 31913322]

[33] Dermody, O.; O'Neill, B.F.; Zangerl, A.R.; Berenbaum, M.R.; DeLucia, E.H. Effects of elevated CO_2 and O_3 on leaf damage and insect abundance in a soybean agroecosystem. *Arthropod-Plant Interact.,* **2008**, *2*(3), 125-135.
[http://dx.doi.org/10.1007/s11829-008-9045-4]

[34] Srinivasa Rao, M.; Manimanjari, D.; Vanaja, M.; Rama Rao, C.A.; Srinivas, K.; Rao, V.U.M.; Venkateswarlu, B.; Jay, R. Impact of elevated CO_2 on tobacco caterpillar, Spodoptera litura on peanut, Arachis hypogea. *J. Insect Sci.,* **2012**, *12*(1), 103.
[PMID: 23437971]

[35] Asala, SW; Olubusuyi, PT; Abdulrazaq, S Evaluation of insect pests associated with growth of soybean *[Glycine max (L.) Merr]* in Nigerian Guinea Savanna. *Int.J. Adv.Agricul.Res,* **2016**, *4*, 51-56.

[36]   Schmitt, VL; Rice, ME; Krell, RK; Van Dee, K *Effects of bean leaf beetle management on soybean yield and on incidence of bean pod mottle virus in Eastern Iowa*; Iowa state research farm progress reports: Eastern Iowa, **2002**, p. 1686.

[37]   Reisig, D. **2020**.*NC State Extension.* Available from: https://content.ces.ncsu.edu/bean-leaf-beetle-in-soybean (Accessed 17 August 2021).

[38]   Anon., *DIGITRAC AGRI BLOG.* **2019**. Available from: https://www.digitrac.in/agricare-blog/how--o-control-tobacco-caterpillar-in-soybean (Accessed 14 August 2021).

[39]   Begna, S.H.; Fielding, D.J. Damage potential of grasshoppers *(Orthoptera: Acrididae)* on early growth stages of small-grains and canola under subarctic conditions. *J. Econ. Entomol.,* **2003**, *96*(4), 1193-1200.
[http://dx.doi.org/10.1093/jee/96.4.1193] [PMID: 14503591]

[40]   Brahman, S.K.; Awasthi, A.K.; Singh, S. Studies on insect pests of soybean (*Glycine max*) with special reference to seasonal incidence of lepidopteran defoliators. *J. Pharmacogn. Phytochem.,* **2018**, *7*(1), 1808-1811.

[41]   Vieira, S.S.; Bueno, R.C.O.F.; Bueno, A.F.; Boff, M.I.C.; Gobbi, A.L. Different timing of whitefly control and soybean yield. *Cienc. Rural,* **2013**, *43*(2), 247-253.
[http://dx.doi.org/10.1590/S0103-84782013000200009]

[42]   Singh, D.P.; Patel, D.U.P. Chemical control pests of soybean. *Indian J. Appl. Res.,* **2011**, *3*(7), 644.
[http://dx.doi.org/10.15373/2249555X/JULY2013/203]

[43]   Jardine,   D.J.   Soybean   Research   &   Information   Network.   **2020**.   Available   from: https://soybeanresearchinfo.com/soybean-pest/bean-leaf-beetle/ (Accessed 14 August 2021).

[44]   Cook,   K.A.   University   of   illinois   IPM.   **2004**.   Available   from: https://ipm.illinois.edu/fieldcrops/insects/corn earworm.pdf (Accessed 17 August 2021).

[45]   Adamu, R.S.; Dike, M.C.; Banwo, O.O. Stinkbug complex of soybean *glycine max* (l.) Merrill and the importance of insecticidal control in the northern guinea savanna area of Nigeria. *Arch. Phytopathol. Pflanzenschutz,* **2001**, *33*(6), 481-489.
[http://dx.doi.org/10.1080/03235400109383370]

[46]   Srinivasan, R.; Tamò, M.; Lee, S.T.; Lin, M.Y.; Huang, C.C.; Hsu, Y.C. *Towards developing a biological control program for legume pod borer, Maruca vitrata. Grain Legumes: Genetic Improvement*; Management and Trade, **2009**, pp. 183-196.

[47]   Nazir, T.; Khan, S.; Qiu, D. Biological control of insect pest. In: *Pests Control and Acarology*; intechopen, **2019**.
[http://dx.doi.org/10.5772/intechopen.81431]

[48]   Gangwar, P.; Trivedi, M.; Tiwari, R.K. Entomopathogenic Bacteria. In: *Microbial Approaches for Insect Pest Management*; Springer: Singapore, **2021**; pp. 59-79.
[http://dx.doi.org/10.1007/978-981-16-3595-3_2]

[49]   Pathak, D.V.; Yadav, R.; Kumar, M. Microbial pesticides: Development, prospects and popularization in India. In: *Plant-Microbe Interactions in Agro-Ecological Perspectives*; Singh, D.; Singh, H.; Prabha, R., Eds.; Springer: Singapore, **2017**; pp. 455-471.
[http://dx.doi.org/10.1007/978-981-10-6593-4_18]

[50]   Pierce, A.A.; Milewski, E.A.; Wozniak, C.A. Federal regulation of plant-incorporated protectants in the United States: Implications for use of bioengineered pesticides in forest restoration. *New For.,* **2022**, 1-11.

[51]   Mehrotra, S.; Kumar, S.; Zahid, M.; Garg, M. In principles and applications of environmental biotechnology for a sustainable future. In: *Biopesticides*; Springer: Singapore, **2017**; pp. 273-292.

[52]   Kumari, P.; Jasrotia, P.; Kumar, D.; Kashyap, P.L.; Kumar, S.; Mishra, C.N.; Kumar, S.; Singh, G.P. Biotechnological approaches for host plant resistance to insect pests. *Front. Genet.,* **2022**, *13*, 914029.

[http://dx.doi.org/10.3389/fgene.2022.914029] [PMID: 35719377]

[53]   Panwar, B.S.; Ram, C.; Narula, R.K.; Kaur, S. Pool deconvolution approach for high-throughput gene mining from *Bacillus thuringiensis*. *Appl. Microbiol. Biotechnol.,* **2018**, *102*(3), 1467-1482.
[http://dx.doi.org/10.1007/s00253-017-8633-6] [PMID: 29177935]

[54]   Talakayala, A.; Katta, S.; Garladinne, M. Genetic engineering of crops for insect resistance: An overview. *J. Biosci.,* **2020**, *45*(1), 114.
[http://dx.doi.org/10.1007/s12038-020-00081-y] [PMID: 33051408]

[55]   Li, H.; Guan, R.; Guo, H.; Miao, X. New insights into an RNAi approach for plant defence against piercing-sucking and stem-borer insect pests. *Plant Cell Environ.,* **2015**, *38*(11), 2277-2285.
[http://dx.doi.org/10.1111/pce.12546] [PMID: 25828885]

[56]   Meng, F.; Li, Y.; Zang, Z.; Li, N.; Ran, R.; Cao, Y.; Li, T.; Zhou, Q.; Li, W. Expression of the double-stranded RNA of the soybean pod borer *Leguminivora glycinivorella* (Lepidoptera: Tortricidae) ribosomal protein *P0* gene enhances the resistance of transgenic soybean plants. *Pest Manag. Sci.,* **2017**, *73*(12), 2447-2455.
[http://dx.doi.org/10.1002/ps.4637] [PMID: 28598538]

[57]   Wolt, J.D.; Wang, K.; Yang, B. The regulatory status of genome-edited crops. *Plant Biotechnol. J.,* **2016**, *14*(2), 510-518.
[http://dx.doi.org/10.1111/pbi.12444] [PMID: 26251102]

[58]   Xu, Y.; Li, Z. CRISPR-Cas systems: Overview, innovations and applications in human disease research and gene therapy. *Comput. Struct. Biotechnol. J.,* **2020**, *18*, 2401-2415.
[http://dx.doi.org/10.1016/j.csbj.2020.08.031] [PMID: 33005303]

CHAPTER 8

Sustainable Crop Nutrition for Ameliorating Biotic Stress in Grain Legumes and Ensuring Food Security

Sifau A. Adejumo[1,*]

[1] *Department of Crop Protection and Environmental Biology, Faculty of Agriculture, University of Ibadan, Ibadan, Nigeria*

Abstract: Environmental stress generally causes considerable yield loss in leguminous crop production. This stress could be biotic (Insect pests, disease pathogens, weeds, vertebrate pests, *etc.*) or abiotic (Drought, heat, cold, salinity, flooding, heavy metal contamination, *etc.*). Either biotic or abiotic stress, both are capable of causing total yield loss. Unfortunately, crops are simultaneously exposed to these stress factors on the field. The response and level of tolerance to both stress factors, however, depend on the crop's genetic and nutritional status. The level of infection or infestation is determined by the cropping system and soil nutrient status. The induction of defense mechanisms by plants in response to pathogenic attack is dependent on environmental conditions like plant nutrient status. It means that there is a complex signaling network with crop nutrition that enables the plants to recognize and protect themselves against pathogens and other environmental stresses. The disease severity could be reduced by adequate crop nutrition due to host nutrient availability, plant composition of secondary metabolites, and the effect on the plant defense mechanisms. Shortages in essential nutrients on their own can predispose plants to attack by pests and pathogens. Therefore, the only sustainable method for growing crops in the face of different environmental stresses is good crop nutrition. A well-fed crop is more resistant to environmental hazards than poorly-fed crop. Though leguminous crops can fix atmospheric nitrogen themselves, the nutritional requirements for healthy crop production are more than just one element. The ability to fix nitrogen, if combined with appropriate crop nutrition will place the plant in a better position to withstand environmental stresses. This chapter discusses some of the different nutrient elements required by leguminous crops and their functions, crop nutrition abiotic stress tolerance, and mechanisms of nutrient-induced resistance in leguminous crops.

Keywords: Fertilisers, Legumes, Environmental Stress, Crop health, Crop yield.

* **Corresponding author Sifau A. Adejumo:** Department of Crop Protection and Environmental Biology, Faculty of Agriculture, University of Ibadan, Ibadan, Nigeria; Tel: +234-803-413-0018; E-mail: sifauadejumo@gmail.com

Phetole Mangena & Sifau A. Adejumo (Eds.)
All rights reserved-© 2023 Bentham Science Publishers

INTRODUCTION

In grain legume production, biotic stress factors such as insect pest attack and pathological diseases have been reported to be the important constraints limiting grain yield [1]. To increase grain yield in the face of biotic stress, different strategies have been proposed. The most important strategy is adequate plant nutrition. A balanced nutrient supply is a basic requirement to protect plants against all forms of stress [2]. The plant growth rate is proportional to nutrient availability and accessibility. A decline in soil fertility has been found to increase the negative crop response and susceptibility to both biotic and abiotic stress. Poor nutrition impairs crop response and tolerance to stress factors. The low grain yields in legumes have been attributed to poor crop management practices and poor soil fertility [1]. Liebig's "Law of the Minimum (1855) stated that 'The genetically fixed yield potential of crops is limited by the nutrition' [2]. The yield potential of any crop is, therefore, determined by the amount of nutrients supplied and taken up by the plant. The presence and availability of essential mineral elements in the soil, therefore, have a significant impact on the plant's health and determine the plant's response to environmental stresses.

Meanwhile, most farmers do not apply additional nutrients to sole cowpea production due to its ability to fix atmospheric nitrogen. But for greater resistance and enhanced tolerance to environmental stress, the addition of fertilizers is needed to boost cowpea tolerance. In fact, nitrogen itself is needed as a starter dose in areas where soils are poor in nitrogen before nodules begin to fix atmospheric nitrogen [3]. Though, nodulation and N fixation can be inhibited by high field N levels due to the inhibition of nitrogenase activity through a feedback mechanism, but moderate/optimal soil nitrogen level is required for effective nodulation [4]. Besides, in the absence of other nutrients like phosphorous, which is critical to cowpea yield, nitrogen fixation is also strongly affected [5]. Phosphorus is the most limiting soil fertility factor for cowpea production in many tropical soils because it stimulates growth, initiates nodule formation, and promotes rhizobium-legume symbiosis apart from other benefits. It means that cowpea nitrogen-fixing ability might also be affected under P deficiency. Again, it has been observed that under stress, the physiological mineral nutrient demand is always higher than that of normal growth. More carbon and nutrients are needed to be able to carry out the stress-induced metabolic activities and ameliorative processes.

The fixed nitrogen might therefore not be able to support leguminous crops under biotic stress. Appropriate and sufficient fertilization is the key to sustainable crop production, especially under stress. The success of pest attack, though, positively correlates with the plant's nutrient status in some reports [6], but the survival or loss encountered is reduced in a well-fed plant. The plant's nutrient status is related to its capacity to ameliorate the negative impacts caused by stress conditions [3, 7]. Best compensatory performance under biotic stress has been reported under proper nutritional management compared to only pest control [6]. Improvement of phosphorus, nitrogen, potassium (P, N, K) and cation contents in the topsoil has been found to increase cowpea grain yield under biotic stress compared to unfertilized fields with pest control [1]. Malnutrition, therefore, predisposes crops to biotic stress. The rate of recovery is also affected or delayed in the absence of balanced nutrition for crops.

Beneficial mineral nutrients should, however, be able to promote growth and yield under stress and strengthen the natural resistance of plants against abiotic and biotic stresses. Apart from mineral nutrients, water is also an essential component of crop nutrition. Legumes like other crops also require more moisture for N fixation. Water is required to export N products from the nodules to the rest of the plant. In the absence of water, N products build up in the nodule and inhibit further fixation by the nodules. With regard to response to biotic stress, lack of water has also been reported to promote insect attack compared to well-watered plants. For instance, aphid performance was found to be the highest in crops subjected to moderate drought stress [8]. Similarly, extreme moisture stress can inhibit nodule initiation or cause nodule shedding in some legume species. It can also reduce N fixation potential by depriving the nodules of sufficient oxygen for rhizobial respiration. Soil nutrient availability and water status can, therefore, have a strong influence and diverse effects on how legumes respond to biotic stresses. The importance of macro and micro-elements in the performance of leguminous crops and tolerance to biotic stresses are discussed in this chapter.

CROP NUTRITION AND BIOTIC STRESS RESPONSES

There are strong interactions between nutrients and other environmental factors, especially, biotic factors. A balanced nutrient supply is the basic requirement to protect plants against all forms of stress. The importance of individual nutrients for maintaining or promoting plant health and growth has been well documented [7]. The level of crop response to biotic stress is dependent on its nutrient status, the type of nutrient available to such crop, and the quantity. It has been observed that an adequate supply of mineral elements in the growth medium is paramount, for plants to survive under different environmental stresses including biotic stress [7, 9]. The growth and survival of leguminous crops under biotic stress are also dependent on the soil nutrient status and ability to fix atmospheric nitrogen effectively. The increased nutrition enables the plants to repair and compensate for the damage caused by insects or pathogens without a reduction in yield. The plant's growth is retarded, the turgor is reduced, and the level of susceptibility to pests and diseases increases under nutrient stress (Fig. **1**).

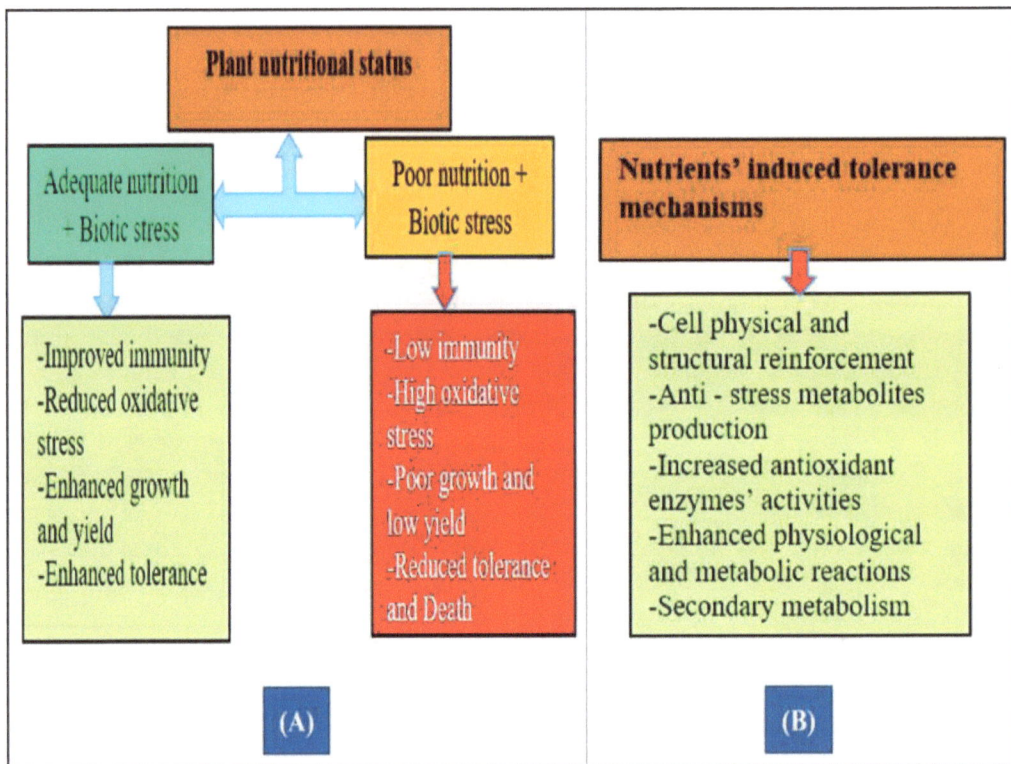

Fig. (1). Crop nutrient status and response to stress (**A**) and the nutrients' induced mechanisms of resistance (**B**).

Nutrient availably or deficiency affects plant-biotic interactions. Plant nutrients, however, function differently in terms of crop response to stress. Some nutrients are antagonistic to disease development, while some are beneficial depending on the nutrient types and concentrations. Lack or presence of nutrients has a direct correlation to pest infestation. For instance, infestation of plants by *Myzus persicae* was not successful in nitrogen-deficient treatment [10]. Meanwhile, the increase in vegetative and reproductive growth in well-nourished plants could serve as attractants to pests and diseases [11]. An increase in the number of flowers because of phosphate treatment was reported to increase the number of cowpea thrips in treated plants compared to the unfertilized ones [6]. Dense canopy formation as a result of nutrient availability has also been reported to provide a suitable micro-environment such as low temperature, high relative humidity, and low light transmission that favoured infestation by *M. vitrata*.

Nutrient's availability therefore has significant effects on the relationship between plant and herbivores because of direct effects of host nutrient availability on the diet of the herbivores and plant's secondary metabolites composition [11]. However, efficient compensatory mechanisms in well-nourished plants would have neutralized the effects of pest infestation. Good nutrition enhances the resistance of cowpeas to insect pests by facilitating the rebuilding of the damaged structures and compensating for the losses caused by the insects. There are different reports on seed priming with various nutrients and water to provide tolerance against biotic stress. Sulphur and silicon-induced resistance against fungal pathogens has also been reported [11, 12]. Silicon for example, though not part of the essential elements, plays a protective role against fungal diseases by positively influencing the structure and function of plant cell walls. It was highly effective against biotrophic and hemi-biotrophic fungi, such as mildew, *Septoria tritici*, and *Fusarium*, after soil application. Cowpea damage by *A. craccivora, M. sjostedti and M. vitrata* was also reduced significantly, and the yield increased with the application of 30 and 45 kg P_2O_5/ha [6].

Mechanisms of Nutrients Induced Resistance

Production of Reactive Oxygen Species (ROS) which consequently leads to oxidative stress is the common effect of environmental stress, either biotic or abiotic. The survival of the plant under oxidative stress depends on its ability to scavenge the ROS as they are being produced. This in turn depends on the plant's nutrient status and vigour. When nutrient deficiency is combined with other environmental stress, the effect is always significant. In well fed plants, the response is quick and the rate of scavenging the ROS is enhanced. Generally, the mechanism for stress alleviation under effective nutrition could be attributed to the activation of antioxidants production in the stressed crop plants and increase in the activities of the antioxidant enzymes. The accumulation of osmoprotectants such as proline (Pro), glycine betaine, glutathione *etc.* is the common physiological response to biotic and abiotic stresses in plants. Stimulation of antioxidative metabolic processes in order to defy oxidative stress is a possible tolerance mechanism being employed under nutrient induced resistance. High nutrition levels have also been reported to alleviate stress damage by sustaining physiological activities like photosynthesis and reduce malondialdehyde (MDA) content under stress [12].

Another important mechanism being induced by sufficient nutrition under stress is the production of secondary metabolites. Several reports indicate the involvement of mineral nutrients in the induction of secondary metabolite synthesis in stressed plants [13, 14]. Carotenoids (Car) pigments for example, are secondary metabolites of isoprenoid origin and are involved in many defense mechanisms, such as membrane stability, light-harvesting, and ROS balance [13]. Synthesis of phytoalexin phenolic compounds has also been reported to be another mechanism employed for Si-induced resistance [2]. The increase in the metabolite production as a result of essential nutrient availability will reduce the success of disease plant interactions. Fortification of the structural integrity of cell walls, the stimulation of the synthesis of defense components, and the contribution to the osmotic adjustment together with ion balance (homeostasis) are also some of the mechanisms being induced for resistance.

Meanwhile, different elements induce different resistance mechanisms in plants under biotic stress. For instance, activities of superoxide dismutase (SOD), peroxidase (POD) and polyphenol oxidase (PPO) have been reported to be enhanced in crops treated with nitrogen ions and exposed to environmental stresses [15]. The glutamine synthetase (GS), and the glutamate synthase (GOGAT) activity increased in all the nitrogen and stress treatments [15]. Silicon has been reported to maintain membrane stability and functions, decrease oxidative damage, and increase antioxidant defence [2, 16]. It increases the structural integrity of cells by incorporating Si (amorphous $(SiO_2)m \cdot n(H_2O)$) in cell walls and intercellular spaces, thereby creating a barrier. Orthosilicic acid is also said to be polymerised to amorphous silica $((SiO_2)m \cdot n(H_2O))$ and deposited in specific cells [2].

Calcium signaling is one of the important responses and immunity under biotic stress [17]. The first crucial step being taken by the plant to adapt to adverse growth conditions is to detect the nature and strength of environmental stimuli, interpret them and activate appropriate physiological responses. In plant pathogen interactions, two types of immune system are triggered. These are Pathogen-Associated Molecular Patterns (PAMP)-Triggered Immunity (PTI) and Effector-Triggered Immunity (ETI). They both enhance overall plant's defense and protects plants from subsequent pathogen attack through localized programmed cell death (PCD) [18]. An increase in calcium concentration during immune signaling is needed for gene reprogramming to initiate adequate responses which could be symbiotic or defensive. Proper response to pathogen attack, therefore, requires high cytosolic calcium ion concentration. The roles of calcium in plant immunity, symbiosis and response to herbivory attacks have been well documented [17, 19, 20].

NUTRIENTS REQUIREMENTS BY LEGUMES AND THEIR FUNCTIONS

Generally, about sixteen nutrient elements are essential for normal plant growth and development. They are grouped under macro and micronutrients. A balanced supply of macro- and micronutrients is essential to decrease the susceptibility of plants to biotic and abiotic stresses. Even though leguminous crops could fix atmospheric nitrogen, other elements must, however, be supplied for normal growth and development because nitrogen cannot substitute for other elements. In the absence of any of the essential elements, the susceptibility of legumes to pest and diseases increases.

Macroelements

Among the essential macronutrients needed by leguminous crops are carbon (C), hydrogen (H), oxygen (O), nitrogen (N), phosphorus (P), potassium (K), calcium (Ca), magnesium (Mg) andsulphur (S). Nitrogen is one of the essential elements that are required by every crop (All the essential elements are required by crops). Without it, the plant might not be able to complete its life cycle and its roles cannot be substituted by another element. It is part of the different biomolecules in plant. It is a structural component of proteins and nucleic acids. Some pigments like chlorophyll also contain nitrogen as their major element. Different physiological and biochemical processes are also enhanced by the presence of nitrogen. Therefore, nitrogen must be supplied to boost crop performance [21]. The main source of nitrogen to the plant is through fertilization except in the leguminous plants where atmospheric nitrogen is fixed by *Rhizobium* and other microorganisms that form nodulation in the roots. As important as nitrogen is to the plants, the optimum requirement also varies for different crops. For instance, the optimum nitrogen level that is tolerable to most leguminous crops is 50 kg N/ha but varies across legume species [22]. This is because the nitrogen fixing ability of legumes has been reported to be hindered by elevated level of nitrogen in the field by disrupting the activity of nitrogenase enzyme through a feedback mechanism and stop nodule formation [23].

It means that nitrogen fixing plant must not be treated with excess nitrogen [23]. Thiourea (TU) is an important synthetic organosulfur compound containing nitrogen (36%) and sulfur (42%) that has gained wide attention for its role in plant stress tolerance [13]. Carbon dioxide is incorporated into the plant through photosynthesis. It is the main source of carbon in plant and carbon is the main energy source for plant. Plant's growth and development are determined by the rate of photosynthesis. It provides the basic skeleton to produce other metabolites in plants. In legumes, photosynthate partitioned to roots supports nodule growth, provides energy for N fixation, maintains a functional population of rhizobia, and allows the synthesis of amino compounds produced from N fixation. The inability of the leguminous crops to accumulate enough carbon through photosynthesis or fertilization has deleterious effects on the growth and plant's ability to withstand stress.

Phosphorous is one of the essential macroelements that perform distinct functions in plants. It is needed for the synthesis of Adenosine Triphosphate (ATP) which is a chemical energy in every living organism including plants. It is a constituent of nucleic acid. Unlike nitrogen, P is not fixed but can only be supplied to legumes in the right quantity. The deficiency has been reported to reduce nodule growth due to the high nodular requirement for P, either directly or indirectly. The P deficiency affects different metabolic activities in crops including legumes. The P deficiency has been attributed to a reduction in nodule formation [24]. The decline in ATP has also been reported under P deficiency. Nitrogen fixation by legumes requires higher P due to a reduction in the nodule formation and energy level. Under stressful environmental conditions, lack of P has been blamed for the low rate of nitrogen fixation. Potassium is also essential for different metabolic processes like photosynthesis, translocation of photosynthates, maintenance of plant turgor, and activation of enzymes. Potassium (K) and sodium (Na) supply are indispensable for osmoregulation and stomatal functioning in plants [25]. Yield losses due to an imbalanced K/Na nutrition were as high as 60% in plants [26].

Magnesium plays a significant role as the central element in the porphyrin ring of chlorophyll. Optimized magnesium concentrations in the nutrient medium are important for maximal CO_2 assimilation rate, as well as for the highest water use efficiency. Stomatal conductance depends on light and magnesium supply. Calcium is one of the essential macronutrients needed for plant growth because of its role in maintaining the structural rigidity of the cell walls as well as in membrane structure and function [27]. Calcium is involved in different physiological processes leading to the growth and development of plants. As discussed earlier, it is also involved in the plant's stress response and categorized as the second important messenger after ROS in stress signalling responses. It serves as a messenger in plant-biotic interactions [17].

Micronutrients

Quantitatively, trace elements are negligible chemical constituents of soils, but are essential as micronutrients for plants. Though, some are not useful biologically, but about seventeen trace elements have now been reported to be useful to plants. According to Bowen [28], the grouping of microelements is based on their activities in plants. For example, those incorporated into structural materials are; Si, Fe, and rarely Ba and Sr, those bound into miscellaneous small molecules, including antibiotics, and porphyrin are As, B, Br, Cu, Co, F, Fe, Hg, I, Se, Si, and V; those combined with large molecules, mainly proteins, including enzymes with catalytic properties are Co, Cr (not certain), Cu, Fe, Mn, Mo, Se, Ni (not certain), and Zn; those fixed by large molecules having storage, transport, or unknown functions are Cd, Co, Cu, Fe, Hg, I, Mn, Ni, Se, and Zn and those related to organelles or their parts (*e.g.,* mitochondria, chloroplasts, some enzyme systems) are Cu, Fe, Mn, Mo, and Zn. Among these, seven are biologically important and essential for proper plant growth and development. These include Fe, Cu, Zn, Bo, Co, Mn, Ni, and possibly Si. The trace elements essential for plants are those that cannot be substituted by others in their specific biochemical roles and that have a direct influence on the plant so that it can neither grow nor complete some metabolic cycle.

In leguminous crops, these micronutrients are also very essential, especially in nodule formation and nitrogen-fixing activities of legumes. The symbiotic relationship between legumes and rhizobials requires micronutrients like B, Co, Zn, Cu, Fe, Mo, Ni, Mn and Se. Apart from the roles of micronutrients in nodulation and activities, they also help in increasing resistance to biotic and abiotic stressors [4]. The micronutrients can achieve this due to their importance in enzyme activation. They are common components of the enzymes. Most micronutrients serve as metal activators in enzymes and by so doing, they enhance the activity of the scavenging and detoxifying enzymes [29]. Iron is an essential microelement with broad/widespread roles in plant's physiological and biochemical processes. It occurs in heme and non-heme proteins. It serves as a metal activator in enzymes and is concentrated mainly in chloroplasts where it is involved in photosynthetic electron transfer. It plays a significant role in chlorophyll synthesis and nucleic acid metabolism. In legumes, Fe is the central element of leghemoglobin.

Molybdenum also functions in the metabolism of plants. It is involved in the nucleic acid synthesis by its function in the enzymatic activities. The nodulated roots have been reported to contain more concentration of Mo compared to other parts of the plants because of their participation in the nitrate reductase activities of the root nodules of leguminous crops. Two Mo-containing enzymes in N metabolism are involved in either nitrogen fixation or nitrate reduction. Manganese is also an essential trace element that is involved in oxidation-reduction reactions and serves as a metal activator in enzymes. It participates in the photosynthetic electron transport system. Mn is also essential in the nodule formation and nitrogen fixation in legumes. Its deficiency and excess affect the rhizobia nodule numbers and thus the efficiency of N fixation. Mn is involved in the nitrite reduction step during nitrogen assimilation.

Nickel is an essential component of the enzyme urease and is required by nodulated legumes to transport N from roots to tops in the forms of ureide compounds. The roles of Co in the plant are generally not clear, although there are some evidence of a favorable effect of Co on plant growth. However, the essentiality of Co for both blue-green algae and microorganisms in fixing nitrogen is now well established. With this, it might also be playing essential roles in leguminous crops. In legumes, Co has been found to affect the ability of plants to fix N_2. It has been reported to be chelated at the center of a porphyrin structure called cobamide coenzyme, which is very important in N_2 fixation. It helps in the transfer of the H atom during the formation of the NH_3 compound by the rhizobia. Its deficiency in the leguminous crop has been reported to inhibit the formation of nodule pigment called ' leghemoglobin' which in turn affects nitrogen fixation. Mo is also a key component of nitrogenase, and its deficiency could disrupt leguminous nodulation and nitrogen fixation.

CONCLUSION

In conclusion, the mineral-nutrient status of plants plays a critical role in increasing plant resistance to environmental stress factors. The nutritional status of plants is relevant to their responses to stress. A balanced supply of macro- and micronutrients is essential to decrease the susceptibility of plants to biotic and abiotic stresses. The resistant varieties developed can only tolerate some levels of attack from these organisms, but the ability is reduced under poor nutrition. The low nutrient status of the plants would reduce the tolerance level of the plant to the attack by reducing the metabolite production. Appropriate fertilization is key. Increasing the physical and chemical fertility of cultivated soils by adequate and balanced supply of mineral nutrients will help in minimizing the detrimental effects of environmental stresses on crop production.

CONSENT FOR PUBLICATION

Not applicable.

CONFLICT OF INTEREST

The author declares no conflict of interest, financial or otherwise.

ACKNOWLEDGEMENT

Declared none.

REFERENCES

[1] Anago, F.N.; Agbangba, E.C.; Oussou, B.T.C.; Dagbenonbakin, G.D.; Amadji, L.G. Cultivation of cowpea; challenges in west Africa for food security: Analysis of factors driving yield gap in Benin. *Agronomy*, **2021**, *11*(6), 1139.
[http://dx.doi.org/10.3390/agronomy11061139]

[2] Haneklaus, S.; Bloem, E.; Schnug, E. Hungry plants : A short treatise on how to feed crops under stress. *Agriculture*, **2018**, *8*(3), 43.
[http://dx.doi.org/10.3390/agriculture8030043]

[3]   Abdul Rahman, N.; Larbi, A.; Kotu, B.; Marthy Tetteh, F.; Hoeschle-Zeledon, I. Does nitrogen matter for legumes? starter nitrogen effects on biological and economic benefits of cowpea (*Vigna unguiculata* L.) in Guinea and Sudan Savanna of West Africa. *Agronomy,* **2018**, *8*(7), 120.
[http://dx.doi.org/10.3390/agronomy8070120]

[4]   Kasper, S.; Christoffersen, B.; Soti, P.; Racelis, A. Abiotic and biotic limitations to nodulation by leguminous cover crops in South Texas. *Agriculture,* **2019**, *9*(10), 209.
[http://dx.doi.org/10.3390/agriculture9100209]

[5]   Nkaa, F.A.; Nwokeocha, O.W.; Ihuoma, O. Effect of phosphorus fertilizer on growth and yield of cowpea (*Vigna unguiculata*). *IOSR J. Pharm. Biol. Sci.,* **2014**, *9*, 74-82.
[http://dx.doi.org/10.9790/3008-09547482]

[6]   Asiwe, J.A.N. The impact of phosphate fertilizer as a pest management tactic in four cowpea varieties. *Afr. J. Biotechnol.,* **2009**, *8*(24), 7182-7186.

[7]   Hasanuzzaman, M.; Fujita, M.; Oku, H.; Nahar, K.; Hawrylak-Nowak, B. *Plant nutrients and abiotic stress tolerance*; Springer Nature Singapore Pte Ltd., **2018**, pp. 1-95.
[http://dx.doi.org/10.1007/978-981-10-9044-8]

[8]   Tariq, M.; Rossiter, J.T.; Wright, D.J.; Staley, J.T. Drought alters interactions between root and foliar herbivores. *Oecologia,* **2013**, *172*(4), 1095-1104.
[http://dx.doi.org/10.1007/s00442-012-2572-9] [PMID: 23292454]

[9]   Feller, U.; Kopriva, S.; Vassileva, V. Plant nutrient dynamics in stressful environments: Needs interfere with burdens. *Agric,* **2018**, *8*, 97.

[10]   Comadira, G.; Rasool, B.; Karpinska, B.; Morris, J.; Verrall, S.R.; Hedley, P.E.; Foyer, C.H.; Hancock, R.D. Nitrogen deficiency in barley *(Hordeum vulgare)* seedlings induces molecular and metabolic adjustments that trigger aphid resistance. *J. Exp. Bot.,* **2015**, *66*(12), 3639-3655.
[http://dx.doi.org/10.1093/jxb/erv276] [PMID: 26038307]

[11]   ur Rehman, H.; Iqbal, Q.; Farooq, M.; Wahid, A.; Afzal, I.; Basra, S.M.A. Sulphur application improves the growth, seed yield and oil quality of canola. *Acta Physiol. Plant.,* **2013**, *35*(10), 2999-3006.
[http://dx.doi.org/10.1007/s11738-013-1331-9]

[12]   Waqas, M.A.; Kaya, C.; Riaz, A.; Farooq, M.; Nawaz, I.; Wilkes, A.; Li, Y. Potential mechanisms of abiotic stress tolerance in crop plants induced by thiourea. *Front. Plant Sci.,* **2019**, *10*, 1336.
[http://dx.doi.org/10.3389/fpls.2019.01336] [PMID: 31736993]

[13]   Uarrota, V.G.; Stefen, D.L.V.; Leolato, L.S.; Gindri, D.M.; Nerling, D. Revisiting carotenoids and their role in plant stress responses: From biosynthesis to plant signaling mechanisms during stress. In: *Antioxidants and Antioxidant Enzymes in Higher Plants*; Gupta, D.; Palma, J.; Corpas, F., Eds.; Springer: Cham, **2018**; pp. 207-232.
[http://dx.doi.org/10.1007/978-3-319-75088-0_10]

[14]   Waqas, M.A.; Kaya, C.; Riaz, A.; Farooq, M.; Nawaz, I.; Wilkes, A.; Li, Y. Potential mechanisms of abiotic stress tolerance in crop plants induced by thiourea. *Front. Plant Sci.,* **2019**, *10*, 1336.
[http://dx.doi.org/10.3389/fpls.2019.01336] [PMID: 31736993]

[15]   Li, S.; Zhou, L.; Addo-Danso, S.D.; Ding, G.; Sun, M.; Wu, S.; Lin, S. Nitrogen supply enhances the physiological resistance of Chinese fir plantlets under polyethylene glycol (PEG)-induced drought stress. *Sci. Rep.,* **2020**, *10*(1), 7509.
[http://dx.doi.org/10.1038/s41598-020-64161-7] [PMID: 32372028]

[16]   Rodrigues, F.Á.; McNally, D.J.; Datnoff, L.E.; Jones, J.B.; Labbé, C.; Benhamou, N.; Menzies, J.G.; Bélanger, R.R. Silicon enhances the accumulation of diterpenoid phytoalexins in rice: A potential mechanism for blast resistance. *Phytopathology,* **2004**, *94*(2), 177-183.
[http://dx.doi.org/10.1094/PHYTO.2004.94.2.177] [PMID: 18943541]

[17]   Aldon, D.; Mbengue, M.; Mazars, C.; Galaud, J.P. Calcium signalling in plant biotic interactions. *Int.*

J. Mol. Sci., **2018**, *19*(3), 665.
[http://dx.doi.org/10.3390/ijms19030665] [PMID: 29495448]

[18]   Jones, J.D.G.; Dangl, J.L. The plant immune system. *Nature,* **2006**, *444*(7117), 323-329.
[http://dx.doi.org/10.1038/nature05286] [PMID: 17108957]

[19]   Hahn, M.G.; Darvill, A.G.; Albersheim, P. Host-pathogen interactions. *Plant Physiol.,* **1981**, *68*(5), 1161-1169.
[http://dx.doi.org/10.1104/pp.68.5.1161] [PMID: 16662068]

[20]   Reddy, A.S.N.; Ali, G.S.; Celesnik, H.; Day, I.S. Coping with stresses: Roles of calcium- and calcium/calmodulin-regulated gene expression. *Plant Cell,* **2011**, *23*(6), 2010-2032.
[http://dx.doi.org/10.1105/tpc.111.084988] [PMID: 21642548]

[21]   Walley, F.L.; Kyei-Boahen, S.; Hnatowich, G.; Stevenson, C. Nitrogen and phosphorus fertility management for desi and kabuli chickpea. *Can. J. Plant Sci.,* **2005**, *85*(1), 73-79.
[http://dx.doi.org/10.4141/P04-039]

[22]   Singh, B.; Usha, K. Nodulation and symbiotic nitrogen fixation of cowpea genotypes as affected by fertilizer nitrogen. *J. Plant Nutr.,* **2003**, *26*(2), 463-473.
[http://dx.doi.org/10.1081/PLN-120017147]

[23]   Graham, P.H. Some problems of nodulation and symbiotic nitrogen fixation in *Phaseolus vulgaris* L.: A review. *Field Crops Res.,* **1981**, *4*, 93-112.
[http://dx.doi.org/10.1016/0378-4290(81)90060-5]

[24]   Carver, T.L.W.; Robbins, M.P.; Thomas, B.J.; Troth, K.; Raistrick, N.; Zeyen, R.J. Silicon deprivation enhances localized autofluorescent responses and phenylalanine ammonia-lyase activity in oat attacked by *Blumeria graminis. Physiol. Mol. Plant Pathol.,* **1998**, *52*(4), 245-257.
[http://dx.doi.org/10.1006/pmpp.1998.0149]

[25]   Zörb, C.; Senbayram, M.; Peiter, E. Potassium in agriculture : Status and perspectives. *J. Plant Physiol.,* **2014**, *171*(9), 656-669.
[http://dx.doi.org/10.1016/j.jplph.2013.08.008] [PMID: 24140002]

[26]   Haneklaus, S.; Knudsen, L.; Schnug, E. Relationship between potassium and sodium in sugar beet. *Commun. Soil Sci. Plant Anal.,* **1998**, *29*(11-14), 1793-1798.
[http://dx.doi.org/10.1080/00103629809370070]

[27]   Hepler, P.K. Calcium: A central regulator of plant growth and development. *Plant Cell,* **2005**, *17*(8), 2142-2155.
[http://dx.doi.org/10.1105/tpc.105.032508] [PMID: 16061961]

[28]   Bowen, H.J.M. *Environmental Chemistry of the Elements*; Academic Press: New York, **1979**, p. 333. 95.

[29]   Rubio, M.C.; Becana, M.; Sato, S.; James, E.K.; Tabata, S.; Spaink, H.P. Characterization of genomic clones and expression analysis of the three types of superoxide dismutases during nodule development in *Lotus japonicus. Mol. Plant Microbe Interact.,* **2007**, *20*(3), 262-275.
[http://dx.doi.org/10.1094/MPMI-20-3-0262] [PMID: 17378429]

CHAPTER 9

Physiological Response of Legumes to Combined Environmental Stress Factors

Ifedolapo O. Adebara[1,*]

[1] *Department of Crop Protection and Environmental Biology, Faculty of Agriculture, University of Ibadan, Ibadan, Nigeria*

Abstract: Legumes are considered the second most important source of food after cereals, and their production can be affected by abiotic and biotic stresses. The incidence of biotic and abiotic stress conditions resulting from climate change is expected to increase in the future and may affect legume production drastically. Abiotic stresses could result in escalated biotic stress occurrence. Although responses to abiotic and biotic stress differ in most cases, combined abiotic and biotic stress responses could be expressed in synergistic or opposing forms. In view of the impending escalation in climate change, responses of legumes to stressful environments are expected to vary among crops. However, collective information on combined biotic and abiotic stress in legumes is not readily available. This paper seeks to gather available information on the responses of legumes to biotic, abiotic, and combined stress with a focus on physiological responses. This review will, therefore, help in providing information and encourage further research into combined stress factors in legumes.

Keywords: Biotic stress, Abiotic stress, Combined stress, Physiological responses, Climate change, Legume production.

INTRODUCTION

Legumes are the largest source of vegetable protein in human diets and livestock feed, they therefore, perform a very important function in reducing protein malnutrition as described by Dita *et al.* [1] and Choudhary *et al.* [2]. Legumes can be either grain or forage, whereby grain legumes include soybean (*Glycine max*), chickpea (*Cicer arietinum*), groundnut (*Arachis hypogaea*), cowpea (*Vigna Unguiculata*), pea (*Pisum sativum*), common bean (*Phaseolus vulgaris*) and pigeon pea (*Cajanus cajan*) amongst others. Meanwhile, forage legumes include alfalfa (*Medicago sativa*), and birds foot trefoil (*Lotus japonicus*), both of which

---

[*] **Corresponding author Ifedolapo O. Adebara:** Department of Crop Protection and Environmental Biology, Faculty of Agriculture, University of Ibadan, Ibadan, Nigeria; Tel: +234 912 357 0150, +234 810 005 8524; E-mail: ifedolapobabalola@gmail.com

Phetole Mangena & Sifau A. Adejumo (Eds.)

have been used as model legume crops for decades. The production of legumes is, however, limited by various biotic and abiotic factors. Abiotic factors, such as drought, extreme temperatures, mineral nutrient imbalances, and salinity are the most important stress factors affecting legumes. Biotic factors such as fungi, viruses, bacteria, nematodes, insects, and weeds are very limiting to the growth and productivity of both grain and forage legumes. Abiotic and biotic stress factors individually severely affect crop yield in general and a combination of these factors can also be detrimental [3].

Many of these biotic and abiotic stress factors are common to all legumes; however, the incidence and severity of these stress vary according to leguminous crop species and the location in which they are grown [1]. These factors manifest in altering physiological activities and metabolism in plants resulting in eventual yield loss of up to 90% or total yield loss depending on the intensity and severity of the stress factor imposed on plants. In the field, these stresses rarely occur in isolation, but they often take place in varying combinations simultaneously [4]. The stress could be in abiotic-abiotic stress combinations or abiotic-biotic stress combinations. Current evidence suggests differences and uniqueness in the plant's ability to respond to a combination of stress as compared to individual stress responses [3]. Furthermore, these stress factors usually affect and influence crop's responses physiologically, morphologically, biochemically, and molecularly, and result in a completely new physiological state in certain cases where the stress has ended, especially when tolerance has been exceeded [5].

Recent predictions have reported expected changes in climatic conditions, predicting sporadic rainfall patterns, warmer temperatures and global warming which will bring about increased incidence of biotic and abiotic stress. Such environmental stress conditions may in turn limit agricultural productivity [6 - 10]. According to Mittler [11], Atkinson and Urwin [12] and Suzuki *et al.* [13] influences of these negative conditions on plants may trigger additive, negative, or interactive effects. In this case, the interactions between various abiotic and biotic stresses may cause significant growth and yield outcomes. On the other hand, abiotic stress can enhance the susceptibility of the crop to pathogen attack while these pathogens may alter the crop's response to abiotic stress factors [3]. Therefore, as this chapter indicates, studying the interactions between stresses and the variations in crop response, particularly to combined stress effects remains pivotal for breeding purposes and developing strategies for stress-tolerant crops.

BIOTIC STRESS IN LEGUMES

Fungal Diseases

The major biotic stress affecting legumes is fungal diseases even though other biotic stresses (viruses, nematodes, insects, bacteria, and weeds) can still result in drastically reduced legume growth and productivity [1]. Fungi, which are biotrophic pathogens cause severe foliar diseases which serve as major constraints in legume production [14]. Fungal foliar diseases that are important affecting legumes are rusts, powdery mildews, and downy mildews. Rust species infecting grain and forage legumes belong to the genus *Uromyces* such as *Uromyces appendiculatus* in common bean, *U. ciceris-arietini* in chickpea, *U. vignae* in cowpea and *U. striatus* mostly infecting alfalfa. Other rust species belonging to other genera that affect legumes include *Phakopsora pachyrhizi* often found in soybeans, *Puccinia arachidis* of groundnut [15] and *Asian rust* that also infects soybean [16]. Other fungal diseases, such as necrotrophic fungal diseases comprise *Ascochyta* blight and *Botrytis* gray mildew which are most common in chickpeas [17]. Soil-borne pathogens known to attack legume crops causing drastic effects in seedlings and adult plants of chickpeas, soybeans, and lentils comprise species of *Fusarium*, *Pythium*, and *Rhizoctonia* [18].

Viral Plant Diseases

Among the many plant pathogens, viral diseases that affect legumes are of particular importance in crop species grown in the subtropical and tropical regions [19]. The viral diseases are transmitted either through seeds or vectors and can also be transmitted by means of mechanical inoculation in cases where induced infections are needed for indexing or research purposes. Legume viruses that are commonly transmitted through seeds include *Bean Common Mosaic Virus* (BCMV), *Cucumber Mosaic Virus* (CMV), *Alfalfa Mosaic Virus* (AMV), *Soybean Mosaic Virus* (SMV), *Peanut Mottle Virus*, *Peanut Stripe Virus*, *Bean Yellow Mosaic Virus* (BYMV) and *Bean Golden Mosaic Virus* (BGMV). These viruses are considered the most limiting viral pathogens in bean production, especially in the Caribbean and also in some parts of Central America. Coyne (year) reported that these viruses resulted in drastic yield losses of up to 100% in many crop fields where they occurred. Other important legume viruses include Groundnut Rosette, which is of high importance in Africa, chickpea stunt occurring both in Africa and Asia [20], pigeon pea sterility mosaic virus, pea bud necrosis virus, and Tobacco streak virus which mainly affects groundnuts [21, 22].

Diseases Caused by Bacteria and Nematodes

Bacterial diseases are of minor importance in legumes when compared to fungi and viruses. However, the most important bacterial diseases infecting these crops are Bacterial blight, Bacterial pustule, and Bacterial wilt caused by *Pseudomonas syringae* p.v *glycinea* in soybean, *Xanthomonas axonopodis* p.v *phaseoli* in common bean, *X.anthomonas campestris* p.v. *cassia* in chickpea, *X. campestris* p.v. *glycine* in soybean and *Pseudomonas solanacearum* in groundnuts [23 - 25]. In addition, nematodes are also not as important as fungal and viral infections. These microbial pathogens infest the crops and show no conspicuous symptoms. Important nematodes are *Heterodera spp.* (Cyst nematode), *Meloidogyne spp.* (Root Knot nematode), *Rotylenchulus spp.* (Reniform nematode), and *Ditylenchus species* (Stem nematodes). Further details on this plant disease-causing microorganisms can be found in reports by Volvas *et al.* [26], Lombardo *et al.* [27], Leach *et al.* [28], and Ahmad and Prasad [29].

Insect Pests

Insects damage crops by acting as vectors for viruses, serving as sources of infection sites, and harming the plants by feeding on vegetative and reproductive parts, as well as fruit pods. Examples of insects that act as vectors include whiteflies, Thrips, Aphids (*Aphis glycine*, *Aphis craccivora*) and Hairy caterpillars [30]. Other insects of economic importance are armyworms (*Spodoptera frugiperda*), green bugs, leaf hoppers, pod borers (*Helicoverpa armigera* and *H. puntigera*) and weevils (*Apion yodmani* and *Zabrodes subfasciatus*).

Weed Plants

Weeds are parasitic plants and pose severe threats to legumes especially grain legume [31]. These undesirable plants affect legumes by competing with them for available mineral nutrients and water, subsequently resulting in major yield reductions of up to 90% depending on the type of weed and intensity of infestation. Infestation of weeds is usually severe in grain legume fields where soybean, pigeon pea, groundnut, chickpea, black gram, and mung bean are cultivated, especially when weed infestations occur in the initial stages of plant growth and development. These parasitic weeds are of various species, which include Striga, Alectra, Avena, Convolvulus, Sonchus, Phaleria, Chenopodium and Orobanche [32, 33]. Striga and Alectra affect grain legumes in the Semi-Arid areas of sub-Saharan Africa, while distinct species of Orobanche are limited in the Mediterranean area, Middle East, Asia, and the United States of America [31].

PHYSIOLOGICAL EFFECTS OF BIOTIC STRESS ON LEGUMES

In grain legumes, biotic stress contributes to poor productivity and yield, and this depends on the type, stage, and severity of infection. For instance, soybean rust affects plants at the flowering to pod-filling stage resulting in 30-100% yield losses, while SMV affects the plants at the late vegetative stage resulting in about 25-50% yield loss [34]. In chickpea, *Ascochyta* blight infects plants at any growth stage, however, leguminous crop plants are more susceptible at the flowering and podding stages which also causes poor seed quality and substantial yield losses [35]. Biotic stress affects the photosynthetic activity of plants in many ways, it disrupts photosynthetic responses either by altering the metabolism of the plants and gas diffusion or by reducing the photosynthetic leaf area. Viruses are, particularly, known to depress the rate of photosynthesis per unit leaf area while fungal infections behave like herbivores in their ability to reduce leaf area, compromise water transport, and induce stomatal closure to reduce photosynthesis as reported by Baron *et al.* [36]. In cowpea, CMV causes reductions in chlorophyll content, leading to significant decreases in the photosynthetic ability, as well as delays on flowering in all infected crop plants.

Nematodes cause a delay in the development of normal growth characteristics, frequently leading to growth retardation in the crops, and preventing legume crops from fixing atmospheric nitrogen with the help of nitrogen-fixing bacteria. The pea cyst nematodes were reported by van Dam and Bouwmeester [37] to cause tissue damage in pea and broad bean plants. Stem nematodes were reported to destroy stem tissues of broad bean plants [26]. An important effect of nematode infection on crops is that they reduce the availability of nutrients and hinder metabolic processes, thereby resulting in reduced plant growth and yield [38]. Win *et al.* [39] indicated that chickpeas and cowpeas are detrimentally affected by root-knot and reniform nematodes. Ring and stunt nematodes cause morphological damage in groundnut plants by forming root-knot and lesion diseases. According to Eschen *et al.* [40] the induced root-knot and lesions lead to about 12% yield losses annually. Generally, in most plant species, the presence of a variety of microbial pathogens always results in necrotic death of cells and tissues [41].

PLANT RESISTANCE TO BIOTIC STRESS

There have been reports of constant changes in the protein content of legumes in response to biotic stress. Plants endogenously regulate the changes in protein profiles and expression of peptides such as proteases and their inhibitors to regulate attacks by pathogenic/ herbivorous organisms. Rodriguez-Moreno *et al.* [42] reported varied patterns of protein turnover, regulation and altered protein profile in *Phaseolus vulgaris* leaves during rust infections using proteo-genomic analyses. As a result, protein determination and quantification can be a functional tool in determining fungal invasion and reproduction [43]. During fungal attacks on crops, salicylic acid signalling has been implicated in the resistance against many biotrophs. In contrast, jasmonic acid and ethylene signalling have been implicated in the resistance against necrotrophs [44]. However, legume crops, use specialised dermal tissues such as trichomes (leaf hairs) and thicker epidermal cells to protect themselves against such infections. Furthermore, some specialised cells of trichomes can secrete exudates possessing antifungal properties and reduce relative humidity making the external surfaces of the plants unfavourable for fungal spore germination [45]. Exudates secreted by glandular trichomes in chickpeas decreased infection rates by *Ascochyta rabiei* and on the other hand, trichomes that were present on the leaf surface of common beans were reported to favour the growth of *Pseudomonas syringae* which may be due to exudates that favoured the growth of microbes, released substances contains nutrients in addition to secondary metabolites and protein digestive enzymes [44, 45].

ABIOTIC STRESS IN LEGUMES

Various abiotic stress constraints posing severe effects on legumes include drought, salinity, chilling, extreme temperatures, mineral nutrient imbalances and flooding. Most abiotic stresses are interconnected and the imposition of one on the plant can result in the introduction of another or result in the introduction or enhancement of biotic stress. According to Gull *et al.* [46], the expression of abiotic stress in plants can be in the form of malfunction in plant cell homeostasis and ion distribution. Amongst all abiotic stresses, drought is the most damaging form of environmental stress. Drought results from an uneven rainfall distribution or inadequate irrigation, leading to significant reductions in available soil water. Many studies have also proposed that drought will worsen because of the predicted continuous climate change. Although some legumes are mostly moderately drought tolerant, especially, cowpeas, peas and chickpeas, a severe drought imposition will cause a drastic yield loss when imposed at flowering and podding stages [47].

Flooding or waterlogging refers to the presence of excess water in the soil caused by the inability of gravity to drain water as fast as it enters the soil [47, 48]. This stress also has negative impacts on legumes, particularly by affecting germination, seedling emergence, root growth, shoot growth, and plant density, causing up to 80% reduction in plant yield development. Waterlogging stress also causes various microbial diseases by providing adequate moisture for the growth of bacteria and fungi for instance [47]. Rasool *et al.*[49] reported that approximately 12% of irrigated lands are affected by salinity worldwide. In comparison to cereal crops, legumes are more affected by high salinity stress, causing reductions in plant growth and development by limiting germination, plant vigour and yield, especially in semi-arid regions as reported by Toker and Mutlu [47], Rasool *et al.* [49] and Hameed *et al.* [50]. Extreme temperatures (either heat or cold stress) also cause negative influences in legume growths and are regarded as some of the major abiotic stresses in legume production [51].

Crops differ in the level of their responses to differing temperatures. Cool season crop species are sensitive to hot temperatures while warm-season species are sensitive to cold temperatures [52]. Plant damage caused by elevated temperatures is usually irreversible after a certain period of exposure, especially if the stress lasts for longer durations [53]. Low soil fertility and toxicity of heavy metals are other causes of yield losses in legumes, together with all other vegetable and grain crops. In legumes, the deficiency of some elements hinders nitrogen fixation resulting in yield reductions. Valentine *et al.* [54], reported that about 40% of the world's arable land is considered acidic. The acidity of the soil determines the availability of certain nutrients, mineral nutrient level, and severity of phytotoxicity elements found in the soil [55].

PHYSIOLOGICAL EFFECTS OF ABIOTIC STRESS ON LEGUMES

Abiotic stress limits agricultural production and physiological processes in a variety of ways. In legumes, abiotic stress disrupts metabolism, resulting in adverse effects on yield and yield components. Drought affects total biomass, pod number, seed number, seed weight, and seed quality in grain legumes [47, 53, 56 - 58]. Photosynthesis and cell growth are primary physiological processes that are limited by drought [54]. In *Medicago truncatula*, leaf relative water content, net carbon dioxide fixation rate, and biochemical photosynthetic processes were maintained during mild drought stress, however, under severe water stress, ribulose-1,5-bisphosphatase (RuBP) regeneration and rubisco carboxylation efficiency were both decreased [59]. Valentine *et al.* [54] also reported a reduction of soybean yield by 40% under drought stress. Furthermore, the stress also reduced water potential and increased abscisic acid (ABA) content in flowers and pods when imposed at critical developmental stages, particularly for the early reproductive stage, resulting in pod abortion, decreased hexose to sucrose ratio in pods, and compromised seed yield in soybean. ABA is a plant hormone that regulates numerous aspects of plant growth, development, and stress response. ABA-deficient mutants from various plant species are widely reported to display reduced seed dormancy and wilted phenotype, clearly demonstrating that this hormone is crucial for plant growth and development.

The decline in metabolic enzyme activity, xylem translocation, transpiration rate, shoot nitrogen demand, and accumulation of ureides in nodules and shoots were reported in legumes leading to a decrease in symbiotic nitrogen fixation (SNF) [60, 61]. Stimulation of sucrose and total sugars and induction of oxidative damage which affects nodule performance have also been reported [61, 62]. Loss of yield due to terminal drought has been reported to range from 26% to 61%, and the loss was attributed to the reduction in dry matter production and partitioning [63]. Waterlogging or flooding leads to a reduction in endogenous nutrient levels in various parts of the plant [64]. Water logging decreases the fresh and dry weight of roots and shoots, the number of nodules per plant as well as plant nitrogenase activity. Other responses of plants to flooding include premature senescence which results in leaf chlorosis, necrosis, defoliation, cessation of growth, and reduced crop yield [65]. High soil salinity can result in other forms of stress such as osmotic stress, ion toxicity, oxidative stress, and nutritional disorders causing alterations in metabolic processes, membrane disorganization, and reduction in cell growth and development [66, 67].

Under salinity conditions, monoatomic/ mono-cationic sodium (Na^+) is often not transported to the leaves, remaining in the roots, while only chloride ions (Cl^-) get transported to the leaves. This shows a differential transport pathway experienced by plants exposed to this stress involving Na^+ and Cl^-, together with a relationship between ion accumulation and salt tolerance. A similar pattern of Na^+ transport was also observed in the shoots of *Lotus japonicus* by Rubio *et al.* [68]. For chickpeas, salinity stress affects the crop negatively at both vegetative and reproductive phases, with severe effects observed during pod formation [69].

Feverish temperatures greater than 35°C lead to a reduction in anthesis and pod setting in most cool season pulses like chickpeas [70]. Elevated temperature stress causes chlorophyll degradation, flower drop, pod abortion, and reduction in seed size during the pod-filling stage in chickpeas [71, 72]. In legumes, nitrate assimilation has been reported to be affected by heat stress. Heat stress causes reductions in ureides synthesis, levels, and activities of nitrate reductase as well as the glutamate synthase enzyme activity [72, 73]. Nitrogenase activity is also affected by heat stress resulting in accelerated nodule senescence and decreased nodule longevity as described by Hungria and Vargas [74] and Serraj *et al.*, respectively [73]. Generally, legumes grown in tropical and subtropical regions or warm-season legumes are sensitive to cold stress or chilling stress. Cold stress-exposed plants of *Medicago truncatula* demonstrated a two-fold increase in stem and root dry matter, followed by an increase in soluble sugars and proline. This study further showed an overall decrease in total dry matter, leaf dry matter, and specific leaf area [75]. In pigeon peas, photosynthesis was completely inhibited under freezing temperatures, also leading to flower and pod drop, delayed maturity, and subsequent loss in yield [76, 77]. Basu *et al.* [78] indicated that chickpeas were also affected by cold stress, predominantly during foliage growth. Additionally, Pande and Sharma [79] also reported that cold stress in combination with wetter conditions makes chickpeas susceptible to increased incidence of *Ascochyta* blight and Anthracnose.

Mineral deficiency, such as the low phosphorus availability is also a stressful factor, especially in tropical legume-growing soils where both nitrogen and phosphorus are usually at low or medium levels [80]. Furthermore, many legume crops such as chickpeas, lentils, peas, soybeans, and common beans have been reported to be severely affected by iron (Fe) deficiency [81]. The sensitivity of legumes to zinc has also been reported especially in common beans, chickpeas, pea, lentils, and soybeans by both Toker and Mutlu [47] and Alloway [82]. Blaylock [83] also reported delayed pod maturity as a result of zinc deficiency in common beans. According to Singh *et al.* [84] aluminum toxicity is considered to be an important limiting factor in plants growing under acidic soils. Aluminum toxicity causes the induction of reactive oxygen species (ROS), lipid peroxidation, root growth restriction, root injury, seedling mortality, abnormal root branching, and calcium, and magnesium deficiencies in many legumes [2, 55, 85].

PLANT RESISTANCE TO ABIOTIC STRESS

During drought, plants adapt to stress through accumulating metabolites and osmoprotectants such as proline, glycine betaine, and nitric oxide. Water-stressed plants also adapt to this stress by reducing the effects of reactive oxygen species and reactive nitrogen species [86]. Some legumes were genetically engineered to accumulate a high content of proline to increase their tolerance to drought stress [87, 88]. Also, manipulation of trehalose metabolism was shown to be a promising strategy to enhance resistance to imposed water deficit and recovery from severe water deficit [89]. In legumes, broad bean, pea, and chickpea are more drought-tolerant than cowpea, soybean, and pigeon pea because the former exports amides (especially asparagine and glutamine) in the nodules to xylem tissues while the latter exports ureides (allantoin and allantoic acid). They also produce antioxidant metabolites such as proline, phenolic compounds, ascorbic acids, and hydrogen peroxide [48].

Moreover, the formation of aerenchyma and adventitious roots is stimulated during waterlogging in soybean plants. However the expansion of these tissues in the roots of many legumes may impede the radial transport of water during water deficit, and the effects may greatly vary when water stress is coupled with salinity stress [90]. Nodules of common beans and chickpeas were also shown to have a high tolerance to salt stress due to a higher enzymatic antioxidant defense [62].

However, salinity stress reduced the activity of nitrogenase and phosphatase enzymes in broad beans [91]. Generally, during plant development, plants exhibit several responses to heat stress which include the expression of heat-shock proteins, modulated levels of hormones, primary and secondary metabolites, and production of reactive oxygen species [53, 92].

COMBINED BIOTIC AND ABIOTIC STRESS EFFECTS

Various abiotic stresses have been reported to result in resistance or susceptibility of crops to varying diseases. For instance, drought and salinity stress have been reported to cause decreased resistance and increased susceptibility of plants to *Puccinia spp.* (Rust), *Verticillium spp.* (Verticillium wilt), *Fusarium spp.* (Fusarium wilt), *Pythium spp.* (Root rot), and *Erysiphe spp.* (Powdery mildew). These effects frequently depend on the nature of the pathogen, severity, and duration of the disease [3]. Drought increased the development of charcoal root rot in *Phaseolus vulgaris* (Common bean) under laboratory conditions [93]. *Uromyces phaseoli* (causal agent of leaf rust) inhibited stomatal closure as a result of the toxins produced by the pathogen, further indicating that the combined stress of drought and leaf rust can compromise drought tolerance in plants [3].

Vigna unguiculata (cowpea) subjected to combined drought and *Macrophomina phaseolina* (causing charcoal rot and stem blight) experienced a marked reduction in shoot water potential. Similarly, *Phaseola vulgaris* (common bean) exposed to combined drought stress and infection by *M. phaseolina* had a high transpiration rate, decreased water potential, and stomatal resistance [93]. In *Glycine max* (soybean), a combination of charcoal rot and drought stress also resulted in reduced leaf water potential [3]. Under concurrent drought and *M. phaseolina*, increased leaf temperature within the tissues and decreased stomatal resistance were observed in stressed common bean plants [93]. In addition, *Arachis hypogea* (groundnut) which were thinned and unthinned subjected to combined drought stress and infection of *Sclerotinia minor* Jagger (watery mildew and soft rots) showed an increased level of disease severity in the unthinned plants. This was because of the increased relative humidity in the unthinned plants, while the thinned plants had lower canopy humidity, and increased water uptake from the soil resulting in lower disease severity [3].

In foliar pathogens, stomatal closure is usually the first physiological barrier used in the protection of plants against pathogen entry into the tissues. Stomatal closure is also a drought avoidance strategy, as such, pathogen-induced stomatal closure helps the plant with efficient use of water during drought by preventing losses due to transpiration [94]. Meanwhile, drought can increase the incidence of soil-borne pathogens especially plant parasitic nematodes [6]. Abscisic acid is produced in response to abiotic stress, while biotic stress responses are determined by mutual antagonism between salicylic acid, jasmonic acid, and ethylene as previously discussed. In biotic and abiotic stress combinations, abscisic acid acts in synergy or antagonistically with the other hormones as reported by Asselbergh *et al.* [95], Yasuda *et al.* [96] and Atkinson and Urwin [12]. A combination of virus, drought, and heat enhanced the expression of defense genes in plants. Nsa and Kareem [97], Balachandran *et al.* [98], and Kumar *et al.* [99] also observed and reported these effects in cowpea plants subjected to a combined virus and drought stress. Contradicting evidence continues to emerge regarding the combined effects of biotic and abiotic stresses in legume growth and production. Such effects remain regardless of both the detrimental and positive effects that these stresses pose on crops. Pandey *et al.* [3] also reported a combination of salinity and ozone stress that reduced the growth and yield of chickpeas substantially. Furthermore, drought and ozone stress were individually reported to have detrimental effects on the growth and development of alfalfa. However, a combination of these stresses may also result in an increased tolerance of the plants to the stress combinations [100].

CONCLUSION

Environmental stress is inevitable and experienced by all plant crops even under seemingly normal environmental conditions. Plants and field crops have to respond to multiple stress levels throughout their growing periods, and leguminous crops are not an exception. The response of crops to abiotic and biotic stress factors could be expressed in numerous ways, either leading to a positive (where a combination results in better development of the crop) effect or a negative response (where a combination of stress results in a more detrimental state of the crop), depending on the type and form of stress. Biochemical, morphological, and physiological information has been documented on the responses of legumes to simultaneously imposed stresses, however, more extensive research is still required to complement plant breeding for desirable growth characteristics and better yield. In addition, with the continuously changing climate, there is a need to study these stress responses to gather insights used in the developmental strategies for managing and preventing impending stress-induced crop losses and damage.

LIST OF ABBREVIATIONS

ABA   Abscisic acid

AMV   Alfalfa mosaic virus

BYMV   Bean yellow mosaic virus

CMV   Cucumber mosaic virus

SMV   Soybean Mosaic virus

SNF   Symbiotic nitrogen fixation

CONSENT FOR PUBLICATION

Not applicable.

CONFLICT OF INTEREST

The author declares no conflict of interest, financial or otherwise.

ACKNOWLEDGEMENT

Declared none.

REFERENCES

[1]   Dita, M.A.; Rispail, N.; Prats, E.; Rubiales, D.; Singh, K.B. Biotechnology approaches to overcome biotic and abiotic stress constraints in legumes. *Euphytica,* **2006**, *147*(1-2), 1-24.
[http://dx.doi.org/10.1007/s10681-006-6156-9]

[2]   Choudhary, A.; Pandey, P.; Senthil-Kumar, M. Tailored responses to simultaneous drought stress and pathogen infection in plants. In: *Drought Stress Tolerance in Plants*; Hossain, M.; Wani, S.; Senthil-Kumar, S., Eds.; Springer: Cham, **2017**; 1, pp. 427-438.
[http://dx.doi.org/10.1007/978-3-319-28899-4_18]

[3]   Pandey, P.; Ramegowda, V.; Senthil-Kumar, M. Shared and unique responses of plants to multiple individual stresses and stress combinations: Physiological and molecular mechanisms. *Front. Plant Sci.,* **2015**, *6*, 723.
[http://dx.doi.org/10.3389/fpls.2015.00723] [PMID: 26442037]

[4]   Berens, M.L.; Wolinska, K.W.; Spaepen, S.; Ziegler, J.; Nobori, T.; Nair, A.; Krüler, V.; Winkelmüller, T.M.; Wang, Y.; Mine, A.; Becker, D.; Garrido-Oter, R.; Schulze-Lefert, P.; Tsuda, K. Balancing trade-offs between biotic and abiotic stress responses through leaf age-dependent variation in stress hormone cross-talk. *Proc. Natl. Acad. Sci.,* **2019**, *116*(6), 2364-2373.
[http://dx.doi.org/10.1073/pnas.1817233116] [PMID: 30674663]

[5]   Fraire-Velázquez, S.; Rodríguez-Guerra, R.; Sánchez-Calderón, L. Abiotic and biotic stress response crosstalk in plants. In: *Physiological, Biochemical and Genetic Perspectives*; intechopen, **2011**.
[http://dx.doi.org/10.5772/23217]

[6]   Atkinson, N.J.; Lilley, C.J.; Urwin, P.E. Identification of genes involved in the response of Arabidopsis to simultaneous biotic and abiotic stresses. *Plant Physiol.,* **2013**, *162*(4), 2028-2041.
[http://dx.doi.org/10.1104/pp.113.222372] [PMID: 23800991]

[7]   Prasch, C.M.; Sonnewald, U. Simultaneous application of heat, drought, and virus to Arabidopsis plants reveals significant shifts in signaling networks. *Plant Physiol.,* **2013**, *162*(4), 1849-1866.
[http://dx.doi.org/10.1104/pp.113.221044] [PMID: 23753177]

[8]   Pandey, P.; Sinha, R.; Mysore, K.S.; Senthil-Kumar, M. Impact of concurrent drought stress and pathogen infection on plants. In: *Combined stresses in plants Cham*; Mahalingam, R., Ed.; Springer International Publishing, **2015**; pp. 203-222.
[http://dx.doi.org/10.1007/978-3-319-07899-1_10]

[9]   Mahalingam, R. Consideration of combined stress: A crucial paradigm for improving multiple stress tolerance in plants. In: *Combined Stresses in Plants*; Mahalingam, R., Ed.; Springer: Cham, **2015**; pp.

1-25.
[http://dx.doi.org/10.1007/978-3-319-07899-1_1]

[10] Ramegowda, V.; Senthil-Kumar, M. The interactive effects of simultaneous biotic and abiotic stresses on plants: Mechanistic understanding from drought and pathogen combination. *J. Plant Physiol.,* **2015**, *176*, 47-54.
[http://dx.doi.org/10.1016/j.jplph.2014.11.008] [PMID: 25546584]

[11] Mittler, R. Abiotic stress, the field environment and stress combination. *Trends Plant Sci.,* **2006**, *11*(1), 15-19.
[http://dx.doi.org/10.1016/j.tplants.2005.11.002] [PMID: 16359910]

[12] Atkinson, N.J.; Urwin, P.E. The interaction of plant biotic and abiotic stresses: from genes to the field. *J. Exp. Bot.,* **2012**, *63*(10), 3523-3543.
[http://dx.doi.org/10.1093/jxb/ers100] [PMID: 22467407]

[13] Suzuki, N; Rivero, RM; Shulaev, V; Blumwald, E; Mittler, R Abiotic and biotic stress combinations. *New Phytologist Foundation,* wiley,. **2014**, *203*(1), 32-43.
[http://dx.doi.org/10.1111/nph.12797]

[14] Sillero, J.C.; Fondevilla, S.; Davidson, J.; Patto, M.C.V.; Warkentin, T.D.; Thomas, J.; Rubiales, D. Screening techniques and sources of resistance to rusts and mildews in grain legumes. *Euphytica,* **2006**, *147*(1-2), 255-272.
[http://dx.doi.org/10.1007/s10681-006-6544-1]

[15] Rubiales, D.; Emeran, A.A.; Sillero, J.C. Rusts on legumes in Europe and North Africa. Grain Legumes. *Eur. J. Plant Pathol.,* **2002**, *37*, 8-9.

[16] Carmona, M.A.; Gally, M.E.; Lopez, S.E. Asian soybean rust: Incidence, severity, and morphological characterization of *Phakopsora pachyrhizi* (*uredinia* and *telia*) in Argentina. *Plant Dis.,* **2005**, *89*(1), 109-109.
[http://dx.doi.org/10.1094/PD-89-0109B] [PMID: 30795304]

[17] Bretag, T.W.; Ramsey, M. Foliar diseases caused by fungi: *Ascochyta spp.* In: *Compendium of pea disease and pests,* 2nd ed; Kraft, J.M.; Pfleger, P.F., Eds.; APS Press: Brazil, **2011**; pp. 24-28.

[18] Hamwieh, A.; Udupa, S.M.; Choumane, W.; Sarker, A.; Dreyer, F.; Jung, C.; Baum, M. A genetic linkage map of Lens sp. based on microsatellite and AFLP markers and the localization of fusarium vascular wilt resistance. *Theor. Appl. Genet.,* **2005**, *110*(4), 669-677.
[http://dx.doi.org/10.1007/s00122-004-1892-5] [PMID: 15650814]

[19] Sastry, K.S.; Zitter, T.A. Management of virus and viroid diseases of crops in the tropics. In: *Plant virus and viroid diseases in the tropics*; Sastry, K.S.; Zitter, T.A., Eds.; Springer B.V: London, **2014**; 1, pp. 149-480.
[http://dx.doi.org/10.1007/978-94-007-7820-7_2]

[20] Kumar, P.L.; Jones, A.T.; Waliyar, F. Virus diseases of pigeon pea. In: *Vegetable and pulse crops, Characterization, diagnosis and management of plant viruses*; Rao, GP.; Kumar, A.T.; Holguin-Peña, RJ., Eds.; Studium Press LLC, **2008**; 3, pp. 235-258.

[21] Kumar, P.L.; Kumari, S.M.G.; Waliyar, F. Virus disease Studies on survival of *Xanthomonas campestris* pv. *glycines. Indian Phytopathol.,* **2008**, *48*, 180-181.

[22] Mandal, B.; Jain, R.K.; Krishnareddy, M.; Krishna Kumar, N.K.; Ravi, K.S.; Pappu, H.R. Emerging problems of Tospoviruses *(Bunyaviridae)* and their management in the Indian Subcontinent. *Plant Dis.,* **2012**, *96*(4), 468-479.
[http://dx.doi.org/10.1094/PDIS-06-11-0520] [PMID: 30727451]

[23] Baltrus, D.A.; Nishimura, M.T.; Dougherty, K.M.; Biswas, S.; Mukhtar, M.S.; Vicente, J.; Holub, E.B.; Dangl, J.L. The molecular basis of host specialization in bean pathovars of *Pseudomonas syringae. Mol. Plant Microbe Interact.,* **2012**, *25*(7), 877-888.
[http://dx.doi.org/10.1094/MPMI-08-11-0218] [PMID: 22414441]

[24]   Khare, M.; Khare, U. Studies on survival of *Xanthomonas campestris* pv. *glycines*. *Indian Phytopathol.,* **2012**, *48*, 180-181.

[25]   Van der Linden, L.; Bredenkamp, J.; Naidoo, S.; Fouché-Weich, J.; Denby, K.J.; Genin, S.; Marco, Y.; Berger, D.K. Gene-for-gene tolerance to bacterial wilt in Arabidopsis. *Mol. Plant Microbe Interact.,* **2013**, *26*(4), 398-406.
[http://dx.doi.org/10.1094/MPMI-07-12-0188-R] [PMID: 23234403]

[26]   Vovlas, N.; Troccoli, A.; Palomares-Rius, J.E.; De Luca, F.; Liébanas, G.; Landa, B.B.; Subbotin, S.A.; Castillo, P. Ditylenchus *gigas n. sp.* parasitizing broad bean: a new stem nematode singled out from the Ditylenchus dipsaci species complex using a polyphasic approach with molecular phylogeny. *Plant Pathol.,* **2011**, *60*(4), 762-775.
[http://dx.doi.org/10.1111/j.1365-3059.2011.02430.x]

[27]   Lombardo, S.; Colombo, A.; Rapisarda, C. Severe damage caused by *Meloidogyne artiellia* on cereals and leguminous in Sicily. *Redia,* **2011**, *94*, 149-151.

[28]   Leach, M.; Agudelo, P.; Lawton-Rauh, A. Genetic variability of *Rotylenchulus reniformis. Plant Dis.,* **2012**, *96*(1), 30-36.
[http://dx.doi.org/10.1094/PDIS-02-11-0132] [PMID: 30731848]

[29]   Koyro, H-W.; Ahmad, P.; Geissler, N. Abiotic stress response in plants. In: *Environmental adaptations and stress tolerance in plants in the era of climate change*; Ahmad, P.; Prasad, M.N.V., Eds.; Springer Cham: United States, **2012**; pp. 1-28.
[http://dx.doi.org/10.1007/978-1-4614-0815-4_1]

[30]   Xu, R.; Li, W.; Zhang, L.F.; Lin, Y.H.; Qi, B.; Xing, H. A study on the inheritance of resistance to whitefly in soybean. *Sci Agric Sin ,* **2010**, *43*, 72-78.

[31]   Rubiales, D.; Pérez-de-Luque, A.; Fernández-Aparico, M.; Sillero, J.C.; Román, B.; Kharrat, M.; Khalil, S.; Joel, D.M.; Riches, C. Screening techniques and sources of resistance against parasitic weeds in grain legumes. *Euphytica,* **2006**, *147*(1-2), 187-199.
[http://dx.doi.org/10.1007/s10681-006-7399-1]

[32]   Doring, T.F.; Storkey, J.; Baddeley, J.; Crowley, O.; Howlett, S.; Mccalman, H.; Pearce, H.; Roderick, S.; Jones, H. Legume based plant mixtures for delivery of multiple ecosystem services: weed diversity and weed control. *SAC SEPA biennial conference: Valuing ecosystems: Policy, economic and management interactions,,* Edinburgh 2012, pp. 163-168.

[33]   Smitchger, J.A.; Burke, I.C.; Yenish, J.P. The critical period of weed control in lentil *(Lens culinaris)* in the pacific Northwest. *Weed Sci.,* **2012**, *60*(1), 81-85.
[http://dx.doi.org/10.1614/WS-D-11-00069.1]

[34]   Coyne, D.P.; Steadman, J.R.; Godoy-Lutz, G.; Gilbertson, R.; Arnaud-Santana, E.; Beaver, J.S.; Myers, J.R. Contributions of the Bean/Cowpea CRSP to management of bean diseases. *Field Crops Res.,* **2003**, *82*(2-3), 155-168.
[http://dx.doi.org/10.1016/S0378-4290(03)00035-2]

[35]   Sharma, H.C.; Srivastava, C.P.; Durairaj, C.; Gowda, C.L.L. Pest management in grain legumes and climate Change. In: *Climate change and management of cool season grain legume crops*; Yadav, S.S.; Redden, R., Eds.; Springer: Dordrecht, **2010**; pp. 115-139.
[http://dx.doi.org/10.1007/978-90-481-3709-1_7]

[36]   Baron, M.; Flexas, J.; Delucia, E.H. Photosynthetic responses to biotic stress. In: *Terrestrial photosynthesis in a changing environment a molecular, physiological and ecological approach*; Flexas, J., Ed.; Cambridge University Press, **2011**; pp. 331-350.
[http://dx.doi.org/10.1017/CBO9781139051477.026]

[37]   van Dam, N.M.; Bouwmeester, H.J. Metabolomics in the rhizosphere: Tapping into belowground chemical communication. *Trends Plant Sci.,* **2016**, *21*(3), 256-265.
[http://dx.doi.org/10.1016/j.tplants.2016.01.008] [PMID: 26832948]

[38] Kerry, B.R. Rhizosphere interactions and the exploitation of microbial agents for the biological control of plant-parasitic nematodes. *Annu. Rev. Phytopathol.,* **2000**, *38*(1), 423-441.
[http://dx.doi.org/10.1146/annurev.phyto.38.1.423] [PMID: 11701849]

[39] Win, P.P.; Kyi, P.P.; Maung, Z.T.Z.; Myint, Y.Y.; Cabasan, M.T.N.; De Waele, D. Host status of rotation crops in Asian rice-based cropping systems to the rice root-knot nematode *Meloidogyne graminicola. Trop. Plant Pathol.,* **2016**, *41*(5), 312-319.
[http://dx.doi.org/10.1007/s40858-016-0106-4]

[40] Eschen, R.; Roques, A.; Santini, A. Taxonomic dissimilarity in patterns of interception and establishment of alien arthropods, nematodes and pathogens affecting woody plants in Europe. *Divers. Distrib.,* **2015**, *21*(1), 36-45.
[http://dx.doi.org/10.1111/ddi.12267]

[41] van Doorn, W.G.; Beers, E.P.; Dangl, J.L.; Franklin-Tong, V.E.; Gallois, P.; Hara-Nishimura, I.; Jones, A.M.; Kawai-Yamada, M.; Lam, E.; Mundy, J.; Mur, L A J.; Petersen, M.; Smertenko, A.; Taliansky, M.; Van Breusegem, F.; Wolpert, T.; Woltering, E.; Zhivotovsky, B.; Bozhkov, P.V. Morphological classification of plant cell deaths. *Cell Death Differ.,* **2011**, *18*(8), 1241-1246.
[http://dx.doi.org/10.1038/cdd.2011.36] [PMID: 21494263]

[42] Rodríguez-Moreno, L.; Barceló-Muñoz, A.; Ramos, C. *In vitro* analysis of the interaction of *Pseudomonas savastanoi* pvs. savastanoi and nerii with micropropagated olive plants. *Phytopathology,* **2008**, *98*(7), 815-822.
[http://dx.doi.org/10.1094/PHYTO-98-7-0815] [PMID: 18943258]

[43] Fernandes, I.; Alves, A.; Correia, A.; Devreese, B.; Esteves, A.C. Secretome analysis identifies potential virulence factors of *Diplodia corticola*, a fungal pathogen involved in cork oak *(Quercus suber)* decline. *Fungal Biol.,* **2014**, *118*(5-6), 516-523.
[http://dx.doi.org/10.1016/j.funbio.2014.04.006] [PMID: 24863480]

[44] Caarls, L.; Pieterse, C.M.J.; Van Wees, S.C.M. How salicylic acid takes transcriptional control over jasmonic acid signaling. *Front. Plant Sci.,* **2015**, *6*, 170.
[http://dx.doi.org/10.3389/fpls.2015.00170] [PMID: 25859250]

[45] Armstrong-Cho, C.; Gossen, B.D. Impact of glandular hair exudates on infection of chickpea by *Ascochyta rabiei. Can. J. Bot.,* **2005**, *83*(1), 22-27.
[http://dx.doi.org/10.1139/b04-147]

[46] Gull, A.; Lone, A.A.; Wani, N.U.I. Biotic and abiotic stresses in plants. In: *Biotic and abiotic stresses in plants*; Gull, A.; Lone, A.A.; Wani, N.U.I., Eds.; IntechOpen: London, **2019**; pp. 1-7.
[http://dx.doi.org/10.5772/intechopen.85832]

[47] Toker, C.; Mutlu, N. Breeding for abiotic stress. In: *Biology and breeding of food legumes*; Pratap, A.; Kumar, J., Eds.; CAB International: United States, **2011**; pp. 241-261.
[http://dx.doi.org/10.1079/9781845937669.0241]

[48] El-Enany, A.E.; Al-Anazi, A.D.; Dief, N.; Al-Taisan, W.A. Role of antioxidant enzymes in amelioration of water deficit and waterlogging stresses on *Vigna sinensis* plants. *J. Biol. Earth Sci.,* **2013**, *3*, B144-B153.

[49] Rasool, S.; Ahmad, A.; Siddiqi, T.O.; Ahmad, P. Changes in growth, lipid peroxidation and some key antioxidant enzymes in chickpea genotypes under salt stress. *Acta Physiol. Plant.,* **2013**, *35*(4), 1039-1050.
[http://dx.doi.org/10.1007/s11738-012-1142-4]

[50] Hameed, A.; Dilfuza, E.; Abd-Allah, E.F.; Hashem, A.; Kumar, A.; Ahmad, P. Salinity stress and arbuscular mycorrhizal symbiosis in plants. In: *Use of microbes for the alleviation of soil stresses*; Miransari, M., Ed.; Springer: New York, United States, **2014**; pp. 139-159.
[http://dx.doi.org/10.1007/978-1-4614-9466-9_7]

[51] Christophe, S.; Jean-Christophe, A.; Annabelle, L.; Alain, O.; Marion, P.; Anne-Sophie, V. Plant N

fluxes and modulation by nitrogen, heat and water stresses: A review. Based on comparison of legumes and non-legume plants. In: *Abiotic stress in plants- mechanisms and adaptations*; Shanker, AK.; Venkateswarlu, B., Eds.; InTechOpen: United Kingdom, London, **2011**; pp. 79-119.

[52]   Hall, A.E. *Crop responses to environment*; CRC Press LLC: Boca Raton, FL, **2001**, pp. 1-24.

[53]   Hasanuzzaman, M.; Nahar, K.; Alam, M.; Roychowdhury, R.; Fujita, M. Physiological, biochemical, and molecular mechanisms of heat stress tolerance in plants. *Int. J. Mol. Sci.,* **2013**, *14*(5), 9643-9684.
       [http://dx.doi.org/10.3390/ijms14059643] [PMID: 23644891]

[54]   Valentine, A.J.; Benedito, V.A.; Kang, Y. Abiotic stress in legume N_2 fixation: From physiology to genomics and beyond. In: *Annual plant reviews: Nitrogen metabolism in plants in the post-genomic era*; Foyer, C.; Zhao, M., Eds.; Wiley-Blackwell: United Kingdom, **2011**; Vol. 42, pp. 207-248.

[55]   Muthukumar, T.; Priyadharsini, P.; Uma, E.; Jaison, S.; Pandey, R.R. Role of arbuscular mycorrhizal fungi in alleviation of acidity stress on plant growth. In: *Use of Microbes for the Alleviation of Soil Stresses*; Miransari, M., Ed.; Springer: New York, United States, **2014**; pp. 43-71.
       [http://dx.doi.org/10.1007/978-1-4614-9466-9_3]

[56]   Khan, H.R.; Paull, J.G.; Siddique, K.H.M.; Stoddard, F.L. Faba bean breeding for drought-affected environments: A physiological and agronomic perspective. *Field Crops Res.,* **2010**, *115*(3), 279-286.
       [http://dx.doi.org/10.1016/j.fcr.2009.09.003]

[57]   Impa, S.M.; Nadaradjan, S.; Jagadish, S.V.K. Drought stress induced reactive oxygen species and antioxidants in plants. In: *Abiotic stress responses in plants: Metabolism, productivity and sustainability*; Ahmad, P.; Prasad, M.N.V., Eds.; Springer: New York, United States, **2012**; pp. 131-147.
       [http://dx.doi.org/10.1007/978-1-4614-0634-1_7]

[58]   Pagano, M.C. Drought stress and mycorrhizal plants. In: *Use of microbes for the alleviation of soil stresses*; Miransari, M., Ed.; Springer: New York, United States, **2014**; pp. 97-110.
       [http://dx.doi.org/10.1007/978-1-4614-9466-9_5]

[59]   Flexas, J.; Medrano, H. Drought-inhibition of photosynthesis in C3 plants: Stomatal and non-stomatal limitations revisited. *Ann. Bot.,* **2002**, *89*(2), 183-189.
       [http://dx.doi.org/10.1093/aob/mcf027] [PMID: 12099349]

[60]   Ahmad, M; Ali, Q; Hafeez, MM; Malik, A Improvement for biotic and abiotic stress tolerance in crop plants. *Biolog. Clin. Sci. Res. J.,* **2021**, 1.
       [http://dx.doi.org/10.54112/bcsrj.v2021i1.50]

[61]   Pandey, P.; Irulappan, V.; Bagavathiannan, M.V.; Senthil-Kumar, M. Impact of combined abiotic and biotic stresses on plant growth and avenues for crop improvement by exploiting physio-morphological traits. *Front. Plant Sci.,* **2017**, *8*, 537.
       [http://dx.doi.org/10.3389/fpls.2017.00537] [PMID: 28458674]

[62]   Arrese-Igor, C; Gordon, C; González, EM; Marino, D; Ladrera, R; Larrainzer, E; Gil-Quintana, E Physiological response of legume nodules to drought. *Plant Stress,* **2011**, *5*(1), 24-31.

[63]   Krishnamurthy, L.; Kashiwagi, J.; Gaur, P.M.; Upadhyaya, H.D.; Vadez, V. Sources of tolerance to terminal drought in the chickpea *(Cicer arietinum L.)* minicore germplasm. *Field Crops Res.,* **2010**, *119*(2-3), 322-330.
       [http://dx.doi.org/10.1016/j.fcr.2010.08.002]

[64]   Ashraf, M.Y.; Ashraf, M.; Arshad, M. Major nutrients supply in legume crops under stress environments. In: *Climate change and management of cool season grain legume crops*; Springer: Dordrecht, **2010**; pp. 155-169.
       [http://dx.doi.org/10.1007/978-90-481-3709-1_9]

[65]   Oosterhuis, D.M.; Scott, H.D.; Hampton, R.E.; Wullschleger, S.D. Physiological responses of two soybean *[Glycine max (L.) Merr]* cultivars to short-term flooding. *Environ. Exp. Bot.,* **1990**, *30*(1), 85-92.

[http://dx.doi.org/10.1016/0098-8472(90)90012-S]

[66]   Conde, A.; Chaves, M.M.; Gerós, H. Membrane transport, sensing and signaling in plant adaptation to environmental stress. *Plant Cell Physiol.,* **2011**, *52*(9), 1583-1602.
[http://dx.doi.org/10.1093/pcp/pcr107] [PMID: 21828102]

[67]   Gürsoy, M.; Balkan, A.; Ulukan, H. Ecophysiological responses to stresses in plants: A general approach. *Pak. J. Biol. Sci.,* **2012**, *15*(11), 506-516.
[http://dx.doi.org/10.3923/pjbs.2012.506.516] [PMID: 24191624]

[68]   Rubio, M.C.; Bustos-Sanmamed, P.; Clemente, M.R.; Becana, M. Effects of salt stress on the expression of antioxidant genes and proteins in the model legume *Lotus japonicus. New Phytol.,* **2009**, *181*(4), 851-859.
[http://dx.doi.org/10.1111/j.1469-8137.2008.02718.x] [PMID: 19140933]

[69]   Samineni, S.; Siddique, K.H.M.; Gaur, P.M.; Colmer, T.D. Salt sensitivity of the vegetative and reproductive stages in chickpea podding is a particular sensitive stage. *Environ. Exp. Bot.,* **2011**, *71*, 260-268.
[http://dx.doi.org/10.1016/j.envexpbot.2010.12.014]

[70]   Kumar, K.; Solanki, S.; Singh, S.N.; Khan, M.A. Abiotic constraints of pulse production in India. In: *Disease of pulse crops and their sustainable management*; Diswas, B.S.K., Ed.; Biotech Books: New Delhi, India, **2016**; pp. 23-39.

[71]   Wang, J.; Gan, Y.T.; Clarke, F.; McDonald, C.L. Response of chickpea yield to high temperature stress during reproductive development. *Crop Sci.,* **2006**, *46*(5), 2171-2178.
[http://dx.doi.org/10.2135/cropsci2006.02.0092]

[72]   du Preez, E.D.; van Rij, N.C.; Lawrance, K.F.; Miles, M.R.; Frederick, R.D. First report of soybean rust caused by *Phakopsora pachyrhizi* on dry beans in South Africa. *Plant Dis.,* **2005**, *89*(2), 206-206.
[http://dx.doi.org/10.1094/PD-89-0206C] [PMID: 30795236]

[73]   Serraj, R.; Sinclair, T.R.; Purcell, L.C. Symbiotic N_2 fixation response to drought. *J. Exp. Bot.,* **1999**, *50*, 143-155.

[74]   Hungria, M.; Vargas, M.A.T. Environmental factors affecting N_2 fixation in grain legumes in the tropics, with an emphasis on Brazil. *Field Crops Res.,* **2000**, *65*(2-3), 151-164.
[http://dx.doi.org/10.1016/S0378-4290(99)00084-2]

[75]   Zhang, L.L.; Zhao, M.G.; Tian, Q.Y.; Zhang, W.H. Comparative studies on tolerance of *Medicago truncatula* and *Medicago falcata* to freezing. *Planta,* **2011**, *234*(3), 445-457.
[http://dx.doi.org/10.1007/s00425-011-1416-x] [PMID: 21523386]

[76]   Rana, D.S.; Dass, A.; Rajanna, G.A.; Kaur, R. Biotic and abiotic stress management in pulses. *Indian J. Agron.,* **2016**, *61*, S238-S248.

[77]   Sultana, R.; Choudhary, A.K.; Pal, A.K.; Saxena, K.B.; Prasad, B.D.; Singh, R.G. Abiotic stresses in major pulses: current status and strategies. In: *Approaches to plant stress and their management*; Gaur, R.K.; Sharma, P., Eds.; Springer: New Delhi, India, **2014**; pp. 173-190.
[http://dx.doi.org/10.1007/978-81-322-1620-9_9]

[78]   Basu, P.S.; Singh, U.; Kumar, A.; Praharaj, C.S.; Shivran, R.K. Climate change and its mitigation strategies in pulses production. *Indian J. Agron.,* **2016**, *61*, S71-S82.

[79]   Pande, M.; Sharma, M. Climate change: potential impact on chickpea and pigeon pea diseases in the rainfed semi-arid tropics (SAT). *Proceedings of the 5th International Food Legumes Research Conference (IFLRC V) and 7th European Conference on Grain Legumes (AEP VII),,* Antalya, Turkey. 2010.

[80]   Srinivasarao, C.H.; Ganeshamurthy, A.N.; Ali, M. *Nutritional constraints in pulse production*; Indian Institute of Pulses Research: Kanpur, India, **2003**.

[81]   Toker, C.; Yildirim, T.; Canci, H.; Inci, N.E.; Ceylan, F.O. Inheritance of resistance to iron deficiency

25

chlorosis in chickpea *(Cicer arietinum L.)*. *J. Plant Nutr.*, **2010**, *33*(9), 1366-1373.
[http://dx.doi.org/10.1080/01904167.2010.484096]

[82]   Alloway, B.J. Soil factors associated with zinc deficiency in crops and humans. *Environ. Geochem. Health*, **2009**, *31*(5), 537-548.
[http://dx.doi.org/10.1007/s10653-009-9255-4] [PMID: 19291414]

[83]   Blaylock, A.D. Navy bean yield and maturity response to nitrogen and zinc. *J. Plant Nutr.*, **1995**, *18*(1), 163-178.
[http://dx.doi.org/10.1080/01904169509364893]

[84]   Singh, D.; Singh, N.P.; Chauhan, S.K.; Singh, P. Developing aluminum-tolerant crop plants using biotechnological tools. *Curr. Sci.*, **2011**, *100*, 1807-1814.

[85]   Singh, D.; Raje, R.S. Genetics of aluminium tolerance in chickpea *(Cicer arietinum)*. *Plant Breed.*, **2011**, *130*(5), 563-568.
[http://dx.doi.org/10.1111/j.1439-0523.2011.01869.x]

[86]   Araújo, S.S.; Beebe, S.; Crespi, M.; Delbreil, B.; González, E.M.; Gruber, V.; Lejeune-Henaut, I.; Link, W.; Monteros, M.J.; Prats, E.; Rao, I.; Vadez, V.; Patto, M.C.V. Abiotic stress responses in legumes: Strategies used to cope with environmental challenges. *Crit. Rev. Plant Sci.*, **2015**, *34*(1-3), 237-280.
[http://dx.doi.org/10.1080/07352689.2014.898450]

[87]   Simon-Sarkadi, L.; Kocsy, G.; Várhegyi, Á.; Galiba, G.; de Ronde, J.A. Genetic manipulation of proline accumulation influences the concentrations of other amino acids in soybean subjected to simultaneous drought and heat stress. *J. Agric. Food Chem.*, **2005**, *53*(19), 7512-7517.
[http://dx.doi.org/10.1021/jf050540l] [PMID: 16159180]

[88]   Kim, G.B.; Nam, Y.W. A novel $\Delta 1$-pyrroline-5-carboxylate synthetase gene of *Medicago truncatula* plays a predominant role in stress-induced proline accumulation during symbiotic nitrogen fixation. *J. Plant Physiol.*, **2013**, *170*(3), 291-302.
[http://dx.doi.org/10.1016/j.jplph.2012.10.004] [PMID: 23158502]

[89]   Duque, A.S.; Almeida, A.; Silva, A.B.; Silva, J.M.; Farinha, A.P.; Santos, D.; Fevereiro, P.; Araujo, S.S. Abiotic stress responses in plants: Unraveling the complexity of genes and networks to survive. In: *Abiotic stress: Plant responses and applications in agriculture*; Vahdati, K.; Leslie, C., Eds.; InTechOpen: London, **2013**; pp. 49-101.

[90]   Pararajasingham, S.; Knievel, D.P. Nitrogenase activity of cowpea *(Vigna unguiculata (L.) Walp.)* during and after drought stress. *Can. J. Plant Sci.*, **1990**, *70*(1), 163-171.
[http://dx.doi.org/10.4141/cjps90-018]

[91]   Hussain, N.; Sarwar, G.; Schmeisky, H.; Al-Rawahy, S.; Ahmad, M. Salinity and drought management in legume crops. In: *Climate change and management of cool season grain legume crops*; Yadav, S.S.; McNeil, D.L.; Redden, R.; Patil, S.A., Eds.; Springer: Dordrecht, **2010**; pp. 171-191.
[http://dx.doi.org/10.1007/978-90-481-3709-1_10]

[92]   Bhattacharya, A. Vijaylaxmi. Physiological responses of grain legumes to stress environments. In: *Climate change and management of cool season grain legume crops*; Yadav, S.S.; McNeil, D.L.; Redden, R.; Patil, S.A., Eds.; Springer: Dordrecht, **2010**; pp. 35-86.
[http://dx.doi.org/10.1007/978-90-481-3709-1_4]

[93]   Mayek-Pérez, N.; GarcÍa-Espinosa, R.; LÓpez-CastaÑeda, C.Á.; Acosta-Gallegos, J.A.; Simpson, J. Water relations, histopathology and growth of common bean *(Phaseolus vulgaris L.)* during pathogenesis of *Macrophomina phaseolina* under drought stress. *Physiol. Mol. Plant Pathol.*, **2002**, *60*(4), 185-195.
[http://dx.doi.org/10.1006/pmpp.2001.0388]

[94]   Tombesi, S.; Nardini, A.; Frioni, T.; Soccolini, M.; Zadra, C.; Farinelli, D.; Poni, S.; Palliotti, A. Stomatal closure is induced by hydraulic signals and maintained by ABA in drought-stressed grapevine. *Sci. Rep.*, **2015**, *5*(1), 12449.

[http://dx.doi.org/10.1038/srep12449] [PMID: 26207993]

[95]   Asselbergh, B.; De Vleesschauwer, D.; Höfte, M. Global switches and fine-tuning-ABA modulates plant pathogen defense. *Mol. Plant Microbe Interact.,* **2008**, *21*(6), 709-719.
[http://dx.doi.org/10.1094/MPMI-21-6-0709] [PMID: 18624635]

[96]   Yasuda, M.; Ishikawa, A.; Jikumaru, Y.; Seki, M.; Umezawa, T.; Asami, T.; Maruyama-Nakashita, A.; Kudo, T.; Shinozaki, K.; Yoshida, S.; Nakashita, H. Antagonistic interaction between systemic acquired resistance and the abscisic acid-mediated abiotic stress response in Arabidopsis. *Plant Cell,* **2008**, *20*(6), 1678-1692.
[http://dx.doi.org/10.1105/tpc.107.054296] [PMID: 18586869]

[97]   Nsa, I.Y.; Kareem, K.T. Additive interactions of unrelated viruses in mixed infections of cowpea *(Vigna unguiculata L. Walp). Front. Plant Sci.,* **2015**, *6*, 812.
[http://dx.doi.org/10.3389/fpls.2015.00812] [PMID: 26483824]

[98]   Balachandran, S.; Hurry, V.M.; Kelley, S.E.; Osmond, C.B.; Robinson, S.A.; Rohozinski, J.; Seaton, G.G.R.; Sims, D.A. Concepts of plant biotic stress. Some insights into the stress physiology of virus-infected plants, from the perspective of photosynthesis. *Physiol. Plant.,* **1997**, *100*(2), 203-213.
[http://dx.doi.org/10.1111/j.1399-3054.1997.tb04776.x]

[99]   Kumar, P.L.; Prasada Rao, R.D.V.J.; Reddy, A.S.; Madhavi, K.J.; Anitha, K.; Waliyar, F. Emergence and spread of Tobacco streak virus menace in Indian and control strategies. *Indian J. Plant Prot.,* **2008**, *36*, 1-8.

[100]   Puckette, M.C.; Weng, H.; Mahalingam, R. Physiological and biochemical responses to acute ozone-induced oxidative stress in *Medicago truncatula. Plant Physiol. Biochem.,* **2007**, *45*(1), 70-79.
[http://dx.doi.org/10.1016/j.plaphy.2006.12.004] [PMID: 17270456]

SUBJECT INDEX

A

B

C